Third International Conference on Science and Technology of Emerging Materials

IOP Conference Series: Materials Science and Engineering
Volume 1286

Pattaya, Thailand
4 – 6 August 2022

Editors:

Wisanu Pecharapa
Sakon Rahong
Wanichaya Mekprasart

ISBN: 978-1-7138-7726-4
ISSN: 1757-8981

Printed from e-media with permission by:

Curran Associates, Inc.
57 Morehouse Lane
Red Hook, NY 12571

Institute of Physics
Dirac House, Temple Back
Bristol BS1 6BE UK

Phone: 44 1 17 929 7481
Fax: 44 1 17 920 0979

techtracking@iop.org

Additional copies of this publication are available from:

Curran Associates, Inc.
57 Morehouse Lane
Red Hook, NY 12571 USA
Phone: 845-758-0400
Fax: 845-758-2633
Email: curran@proceedings.com
Web: www.proceedings.com

TABLE OF CONTENTS

IOP Conf. Series: Materials Science and Engineering 1286 (2023) 011001 doi:10.1088/1757-899X/1286/1/011001

International Conference on Science and Technology of Emerging Materials 2022

STEMa 2022
International Conference on Science and Technology of Emerging Materials

4th-6th August 2022, Pattaya, Thailand

Peer reviewed papers from International Conference on Science and Technology of Emerging Materials 2022

Edited by

Wisanu Pecharapa, Sakon Rahong, Wanichaya Mekprasart

Conference Organized by

CMIT COLLEGE OF MATERIALS INNOVATION AND TECHNOLOGY

College of Materials Innovation and Technology, King Mongkut's Institute of Technology Ladkrabang

IOP Conf. Series: Materials Science and Engineering 1286 (2023) 011001 doi:10.1088/1757-899X/1286/1/011001

PREFACE

The International Conference on Science and Technology of Emerging Materials 2022 (STEMa2022) held in enchanting city of Pattaya, Thailand is aimed to draw attention and participation from over 200 researchers in the field of materials science and related technology. It is also a great event for students, scientists, researchers, technologists to exchange their idea with others in the knowledge of emerging materials and initiate new research network with new international colleague and collaboration.

All articles in this issue have been subjected to a peer-review process based on their originality and quality. The topics covered in this book have covered following topics;

- Advanced Functional Materials
- Biomaterials and Carbon-based Materials
- Computational Materials
- Energy and Environment Materials
- Materials Fabrication, Characterization and Manufacturing
- Materials for Health Science
- Nano-Materials and Applications
- Optical and Electronic Materials
- Sensors, Sensing Materials and Related Devices

The organizing committee would like to express appreciation to all supporters, sponsors and participants for great contribution to the conference success.

STEMa2022 organizing committee

IOP Conf. Series: Materials Science and Engineering 1286 (2023) 011001 doi:10.1088/1757-899X/1286/1/011001

STEMa2022 Committee

Chairman:

Asst. Prof. Dr. Wipoo Sriseubsai	College of Materials Innovation and Technology, KMITL
Assoc. Prof. Dr. Wisanu Pecharapa	College of Materials Innovation and Technology, KMITL

International Advisory Board:

Prof. Dr. Keiichi N. Ishihara	Kyoto University, Japan
Prof. Dr. Hidenori Mimura	Shizuoka University, Japan
Prof. Dr. Yasuhino Hayakawa	Shizuoka University, Japan
Prof. Dr. Toru Aoki	Shizuoka University, Japan
Prof. Dr. Masayoshi Fuji	Nagoya Institute of Technology, Japan
Prof. Dr. Young Jae Song	Sungkyunkwan Univeristy, South Korea
Prof. Dr. C.K. Jayasankar	Sri Venkateswanra University, India
Prof. Dr. Balaji Rao Ravuri	GITAM Deemed to be University, India
Prof. Dr. Shobha Rani Depuru	Institute of Aeronautical Engineering, India
Prof. Dr. Mitra Djama	Institute of Technology Bandung, Indonesia
Prof. Dr. Jürgen Rühe	University of Freiburg – IMTEK, Germany
Assoc. Prof. Dr. Rizalinda de Leon	University of the Philippines, Philippines
Assoc. Prof. Dr. Volodymyr Gnatyuk	National Academy of Science of Ukraine, Ukraine
Assoc. Prof. Dr. Takao Yasui	Nagoya University, Japan
Dr. Taras Kolodiazhnyi	Huawei Technologies Japan, Japan
Dr. Kateryna Zelenska	Taras Shevchenko National University of Kyiv, Ukraine
Dr. Ken Albrecht	Tokyo Institute of Technology, Japan
Dr. Vijayan Manickam Achari	University of Malaya, Malaysia
Dr. Mukannan Arivanandhan	Anna University, India

Local Advisory Board:

Prof. Dr. Jiti Nukeaw	College of Materials Innovation and Technology, KMITL

Local Organizing Committee:

Assoc. Prof. Dr. Apiluck Eiad-Ua	College of Materials Innovation and Technology, KMITL
Assoc. Prof. Dr. Benchapol Tunhoo	College of Materials Innovation and Technology, KMITL
Assoc. Prof. Dr. Navaphun Kayunkid	College of Materials Innovation and Technology, KMITL
Assoc. Prof. Dr. Sutee Chutipaijit	College of Materials Innovation and Technology, KMITL
Assoc. Prof. Dr. Wanwilai Vittayakorn	College of Materials Innovation and Technology, KMITL
Assoc. Prof. Dr. Winadda Wongwiriyapan	College of Materials Innovation and Technology, KMITL

IOP Conf. Series: Materials Science and Engineering 1286 (2023) 011001 doi:10.1088/1757-899X/1286/1/011001

Assoc. Prof. Dr. Tosapol Maluangnont	College of Materials Innovation and Technology, KMITL
Asst. Prof. Dr. Adirek Rangkasikorn	College of Materials Innovation and Technology, KMITL
Asst. Prof. Dr. Anuchit Jaruvanawat	College of Materials Innovation and Technology, KMITL
Asst. Prof. Dr. Kanoknan Phacheerak	College of Materials Innovation and Technology, KMITL
Asst. Prof. Dr. Kanokthip Boonyarattanakalin	College of Materials Innovation and Technology, KMITL
Asst. Prof. Dr. Kittiphong Amnuyswat	College of Materials Innovation and Technology, KMITL
Asst. Prof. Dr. Korakot Onlaor	College of Materials Innovation and Technology, KMITL
Asst. Prof. Dr. Mayuree Phonyiem	College of Materials Innovation and Technology, KMITL
Asst. Prof. Dr. Nongluck Houngkamhang	College of Materials Innovation and Technology, KMITL
Asst. Prof. Dr. Pitiporn Thanomngam	College of Materials Innovation and Technology, KMITL
Asst. Prof. Dr. Sakon Rahong	College of Materials Innovation and Technology, KMITL
Asst. Prof. Dr. Supamas Wirunchit	College of Materials Innovation and Technology, KMITL
Asst. Prof. Dr. Thutiyaporn Thiwawong	College of Materials Innovation and Technology, KMITL
Asst.Prof.Dr. Wanichaya Mekprasart	College of Materials Innovation and Technology, KMITL
Dr. Khattiya Chalapat	College of Materials Innovation and Technology, KMITL

IOP Conf. Series: Materials Science and Engineering 1286 (2023) 011001 doi:10.1088/1757-899X/1286/1/011001

Organized by

Co-Organized by

Sponsors

Peer Review Statement

All papers published in this volume have been reviewed through processes administered by the Editors. Reviews were conducted by expert referees to the professional and scientific standards expected of a proceedings journal published by IOP Publishing.

- **Type of peer review:** Single Anonymous
- **Conference submission management system:** Morressier
- **Number of submissions received:** 12
- **Number of submissions sent for review:** 11
- **Number of submissions accepted:** 10
- **Acceptance Rate (Submissions Accepted / Submissions Received × 100):** 83.3
- **Average number of reviews per paper:** 1
- **Total number of reviewers involved:** 10
- **Contact person for queries:**
  Name: Wisanu Pecharapa
  Email: wisanu.pe@kmitl.ac.th
  Affiliation: King Mongkut's Institute of Technology Ladkrabang - College of Materials Innovation and Technology

IOP Conf. Series: Materials Science and Engineering 1286 (2023) 012001 doi:10.1088/1757-899X/1286/1/012001

Effect of Copper Percentage in Copper Tungsten Electrode on Ti6Al4V by Electrical Discharge Machining

A Muttamara[1], P Nakwong[2] and S Pararach[3]

[1,3] Faculty of Engineering, Thammasat School of Engineering(TSE),
Thammasat University, Phatunthani, 12120, Thailand
[2] Department of Applied Science, Faculty of Science and Technology, Phranakhon Si Ayutthaya Rajabhat University, Thailand
[1]E-mail : mapiwat@engr.tu.ac.th

Abstract. The aim of this study was to identify the optimal parameters for electrical discharge machining of Ti6Al4V titanium alloy using copper tungsten electrodes with tungsten compositions of 70% and 80%, respectively. The experiment involved varying current, pulse on time, and duty factor were determined using the Taguchi experimental method L-18 orthogonal arrays. The level of importance of each machining parameter on material removal rate (MRR) was determined using analysis of variance (ANOVA), and the optimal machining parameter combination was obtained using the analysis of signal-to-noise (S/N) ratio. The results showed that a Cu30/W70% electrode performed the best in terms of MRR, electrode wear, and surface roughness. The optimal machining conditions for achieving maximum MRR were found to be a current of 30 A, a pulse on time of 50 μs, and a duty factor of 40%.

1. Introduction

Electrical Discharge Machining (EDM) is one of non-conventional material removal process which it basically widely used in industry [1]. The performance of the EDM process can be extended for hard materials, depends on the material and the electrode materials [2-3]. Titanium alloys are also regarded as "difficult-to-machine" materials because of their outstanding properties such as a high strength-to-weight ratio, good resistance to creep and fatigue and excellent corrosion resistance. They are increasingly being used in the following areas: aerospace applications, automotive applications, biomedical applications and other industries [4]. Hence, Electrical Discharge Machining (EDM) can be a way out to machine it precisely[4-6]. A combination of the advantages of tungsten and copper, high temperature resistance, electric arc ablation, high intensity, than the major, conductive, thermal conductivity, and ease of machining, and it has features such as cold sweating, as with tungsten high hardness, high melting point, anti-adhesion characteristics, often used to do a certain resistance to abrasion, resistant to high temperature EDM.Tungsten-copper (W-Cu) electrodes have been used widely for machining tungsten carbide and tool steel work-pieces because of the combination of a high thermal and electric conductivity, high strength, high wear resistance, and excellent arc erosion resistance in a low thermal expansion coefficient composite [2–3]. The Taguchi method is a powerful tool for parametric design of performance characteristics, is used to determine the optimal machining parameters for minimum electrode wear ratio, maximum material removal rate and minimum surface roughness in the EDM operations [7-8]. Muttamara and Phimoolchat focused on the application of Taguchi and grey relational analysis with entropy measurement for multiobjective optimization in electrical discharge machining (EDM) of Ti6Al4V. The use of Taguchi and grey relational analysis, along with entropy measurement, allowed for the identification and optimization of multiple objectives simultaneously [9]. The experimental details when using the Taguchi method are described. EDM was compared with W-Cu alloy:70W-30Cu and 80W-20Cu and performed with, at various parameters. The machinability of tungsten carbides was evaluated in terms of material removal rate, electrode wear and surface roughness.

2. Materials and Methods

Titanium alloy grade Ti6Al4V is to be used as work-piece material. Table 1 shows the material related properties. Experiments were performed using a Mitsubishi EA8 Model E8 – ADVANCE SERIES as shown in Figure 1. Two grade of copper tungsten electrode with a diameter of 2 mm were used as electrodes. Table 2 shows material properties of CuW alloy. The MRR, EWR and surface roughness were evaluated for each cutting condition. Machining experiments for determining the optimal machining parameters were carried out by setting negative polarity. The working levels was selected under the preliminary experiment. Table 3 shows working levels. The Taguchi method can be designed according to the L18 Orthogonal Array which shown in table 4.

Table 1. Chemical compositions of a workpiece.

Elements	(wt%)
Al	6
Fe	Max 0.25
O	Max 0.20
Ti	90
V	4

Table 2. Material properties of CuW alloy.

Topic	Cu30/W70	Cu20/W80
Density (g/cm^3)	13.8	15.15
Hardness (HB)	175	220
Resistivity (μΩ.cm)	4.1	5
IACS (%)	42	34
Bending Strength(MPa)	790	980

Table 3. Factors and details including the level used in Taguchi method.

Factors	Level1	Level2	Level3
1.Electrode	Cu30/W70	Cu20/W80	
2.Current (A)	5	20	30
3.On time (μs)	50	100	400
4.Duty Factor (%)	5	40	80

Figure 1. EDM Machine.

Table 4. L18 Orthogonal Array that used in Taguchi method.

No.	Electrode	Current	On time	Duty factor
1	1	1	1	1
2	1	1	2	2
3	1	1	3	3
4	1	2	1	1
5	1	2	2	2
6	1	2	3	3
7	1	3	1	2
8	1	3	2	3
9	1	3	3	1
10	2	1	1	3
11	2	1	2	1
12	2	1	3	2
13	2	2	1	2
14	2	2	2	3
15	2	2	3	1
16	2	3	1	3
17	2	3	2	1
18	2	3	3	2

3. Results and Discussion

3.1. The material removal rate from experiment by using Taguchi method

Larger the better was used for determination of MRR. The experiment was repeated twice (n=2) and Signal-Noise-Ratio(S/N) was calculated by equation (1).

$$S/N_L = -10log(\frac{1}{n}\sum_{i=1}^{n}\frac{1}{y_i^2}) \quad (1)$$

The experiment's results were applied to analyse variance by utilizing Minitab 18.0, providing the average Signal-Noise-Ratio (S/N) as presented in Table 5 and Figure 2. The difference between the maximum and minimum S/N ratio is called delta, and the greater its absolute difference, the more significant the factor to MRR (material removal rate). The ranking of the greatest delta values, with current factor having the highest delta value of 25.826, followed by on-time with a delta of 6.012, duty factor with a delta of 1.75, and electrode with a delta of 0.648, reflects the relative strength of the impact that each factor has on the MRR. The study found that the most significant factors that affect the MRR are current, on-time, duty factor, and electrode, with the highest impact attributed to current followed by on-time, duty factor, and electrode. The high delta value for current indicates that even small variations in current can significantly impact MRR, while the lower delta values for on-time, duty factor, and electrode suggest that changes in these factors have a relatively smaller impact on MRR. Therefore, to optimize MRR, it is essential to first and foremost focus on controlling the current factor, followed by adjusting the other factors accordingly.

Based on the ranking of the most significant factors, the greatest machining condition to achieve maximum MRR was found to be electrode Cu30/W70, current of 30 A at level 3, on-time of 50 μs at level 1, and duty factor of 40% at level 2. Furthermore, the study also highlighted that an increase in current would raise the temperature of electric discharge, leading to superior MRR [10-11]. These findings provide insights for the selection of machining parameters to improve MRR and optimize the electro-discharge machining process.

Table 5. Response for S/N ratio of factors for Material removal rate.

Level	Electrode	Current	On time	DF
1	9.604	-5.785	12.021	10.360
2	10.252	15.528	11.754	10.590
3		20.042	6.009	8.835
Delta	0.648	25.826	6.012	1.755
Rank	4	1	2	3

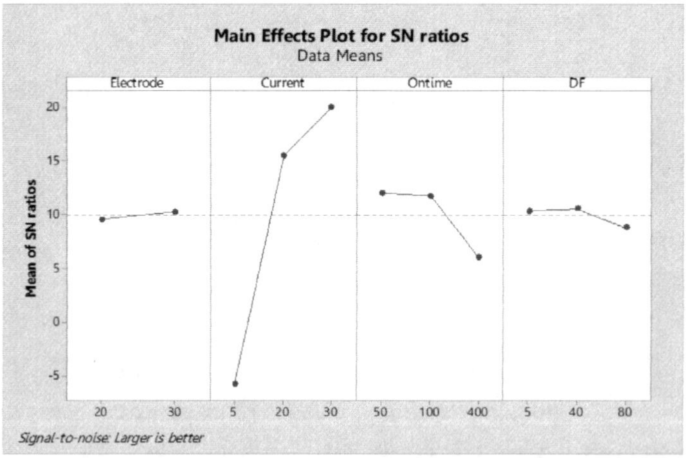

Figure 2. Main effects for Signal-Noise-Ratio (S/N) ratio of each factor on MRR.

3.2. Electrode wear ratio by Taguchi method

Smaller the better was used for determination of EWR. The experiment for electrode wear ratio was repeated twice (n=2) and Signal-Noise-Ratio(S/N) was calculated by equation (2). Apply the result of the experiment to provide the average Signal-Noise-Ratio (S/N) which shows in table 6 and plot graph as shown in figure 3.

$$S/Ns = -10log(\frac{1}{n}\sum_{i=1}^{n} y_i^2) \quad (2)$$

The result that the most significant factor affecting EWR in the experiment is on-time is an interesting finding. On-time represents the duration of the spark discharge during the EDM process, and it is a critical parameter that directly affects the material removal rate and surface quality. The high delta value of on-time indicates that even small changes in on-time can have a significant impact on EWR. This suggests that controlling the on-time parameter is crucial in achieving the minimum EWR in the EDM process. It is worth discussing how on-time affects the EWR. When the on-time is too long, the discharge energy will be transferred to the electrode and not to the workpiece, resulting in excessive electrode wear and increased EWR. On the other hand, when the on-time is too short, the discharge energy may not be sufficient to remove the material efficiently, leading to increased electrode wear and EWR. Figure 4 shows electrode shapes obtained after EDM process; a) Cu30/W70 and b) Cu20/W80. The larger corner radius observed in Cu30/W70 compared to Cu20/W80 can be attributed to the difference in the material properties of the two electrode compositions. Cu30/W70 contains a higher percentage of tungsten (70%), which contributes to its increased hardness and wear resistance. As a result, during the electrical discharge machining (EDM) process, the Cu30/W70 electrode experiences reduced wear and deformation, leading to a smoother cutting action and larger corner radius.

Table 6. Response for S/N ratio of factors for electrode wear ratio.

Level	Electrode	Current	Ontime	DF
1	-7.988	-3.587	-11.503	-10.816
2	-6.550	-13.513	-14.240	-9.065
3		-4.707	3.937	-1.926
Delta	1.438	9.925	18.178	8.890
Rank	4	2	1	3

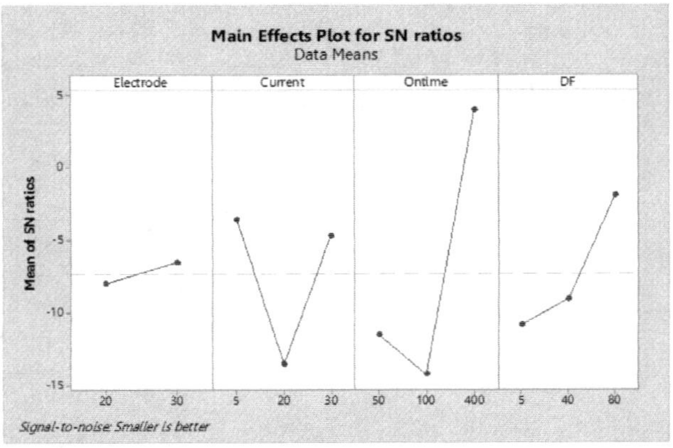

Figure 3. Main effects for Signal-Noise-Ratio (S/N) of each factor on EWR.

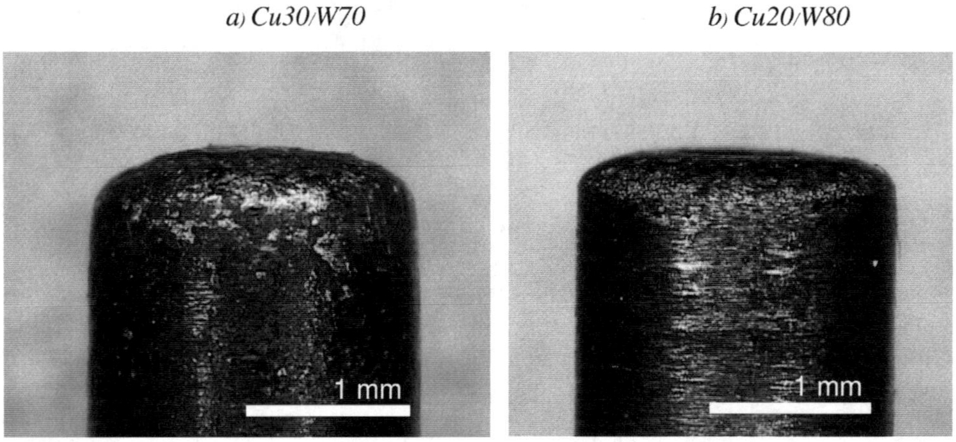

Figure 4. Electrode shapes obtained after EDM process; a) Cu30/W70 and b) Cu20/W80.

3.3. Surface roughness of workpiece by Taguchi method

Smaller the better was used for determination of surface roughness (Ra). The result of the experiment to provide the average Signal-Noise-Ratio (S/N) which shows in figure 5. Table 7 shows Response for S/N ratio of factors for surface roughness. The greater the absolute difference between maximum and minimum S/N ratio the significant the factor to Ra is. Most significant factors are current shows the greatest delta of current is 3.101 followed by on time with delta of 1.783, delta value of duty factor is 1.037 and delta of electrode is 0.460. Surface roughness increases with slightly increases with peak current. Larger craters were produced by a larger producing a larger discharging energy as shown in figure 6 [11]. The optimum machining condition to obtain minimum Ra was found to be current = 5 A at level 1, on time = 100 µs at level 2, duty factor = 5% at level 1 and electrode is CuW20% at level 1.

Table 7. Response for S/N ratio of factors for surface roughness.

Level	Electrode	Current	Ontime	DF
1	-11.213	-9.123	-10.319	-10.300
2	-10.753	-12.224	-10.528	-11.337
3		-11.601	-12.102	-11.312
Delta	0.460	3.101	1.783	1.037
Rank	4	1	2	3

IOP Conf. Series: Materials Science and Engineering 1286 (2023) 012001 doi:10.1088/1757-899X/1286/1/012001

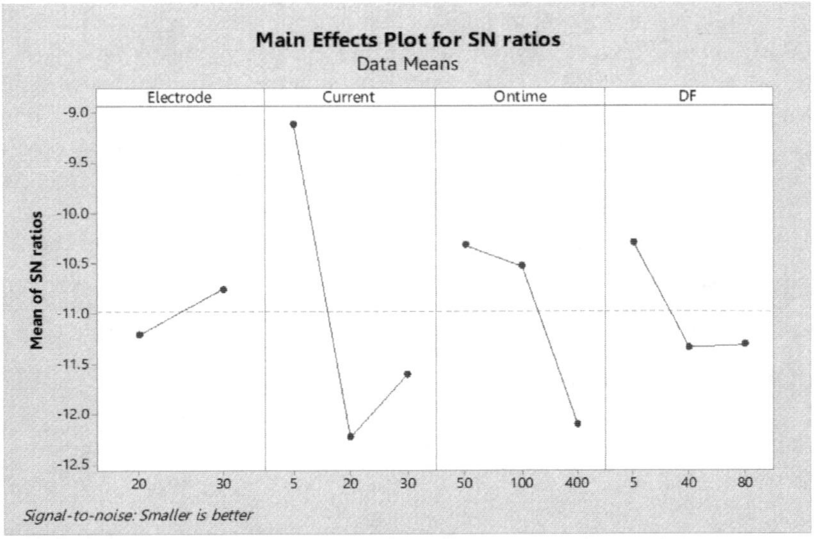

Figure 5. Main effects for Signal-Noise-Ratio (S/N) of each factor on Ra.

a) I=5A *b) I =30A*

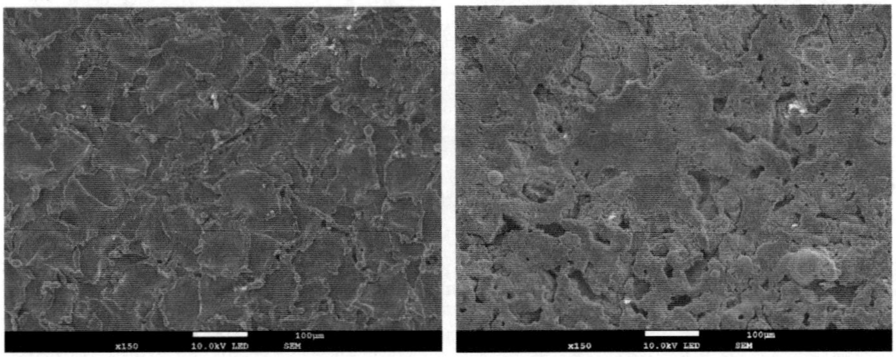

Figure 6. EDMed surface with CuW20%, on time = 100 μs, duty factor = 5% at a) I=5A and b) I =30A.

3.4. Confirmation Test

A confirmation test was conducted to evaluate the quality characteristics for EDM using the optimal conditions ($A_1B_3C_1D_2$). Table 8 compares the experimental results for MRR obtained under these optimal conditions with the predicted results. The predicted MRR value was estimated to be 13.51 mm³/min, while the experimental result gave 13.92 mm³/min. The table shows a difference of 0.41 mm³/min between the experimental and estimated results. The use of CuW electrode with Cu30/W70 resulted in a higher MRR than that of Cu20/W80, indicating that the electrical conductivity of copper in CuW material electrode plays an important role in discharging. The difference between the experimental and estimated results could be attributed to various factors such as variations in the material properties of the workpiece, tool wear, or machine calibration issues.

Table 8. Result of confirmation experiment for MRR.

Response value	**Best Conditions**		**Conditions**	**Difference**
	Estimation	Experiment	Experiment	
Level	$A_2B_3C_1D_2$		$A_1B_3C_1D_2$	
MRR	13.51	13.92	13.85	0.41
S/N Ratio	23.12		22.83	

The comparisons of the EWR according to optimal test and the predicted combination $A_2B_1C_3D_3$, and the experiment $A_1B_1C_3D_3$ are given in table 9. According to these comparisons, The predicted values were obtained by estimation of EWR = -0.78% while as an experimental result gives 0.02%. The difference between experimental result and the estimated result is 0.8. Table 10 shows the result of confirmation Experiment for Ra. The best conditions from taguchi method were compared with conditions from experiment #run no.1 (A1B1C1D1). This is corresponding to discharge energy, lower discharge gives smoother surface roughness due to molten material and debris from the gap is removed a little. According to these comparisons, the predicted values were obtained by estimation of surface roughness (SR) = 2.47 μm while as an experimental result gives 2.68 μm.

Table 9 Result of confirmation experiment for EWR.

Response value	Best Conditions		Conditions	Difference
	Estimation	Experiment	Experiment	
Level	$A_2B_1C_3D_3$		$A_1B_1C_3D_3$	
EWR	-0.78	0.02	0.02	0.8
S/N Ratio	13.68		33.97	

The difference between experimental result and the estimated result is 0.05. CuW electrode with Cu30/W70 gives smoother surface roughness than that of Cu20/W80 due to electrical conductivity property of Cu30/W70 is better than that of Cu20/W80. This might be explained that the area of workpiece is more sparked throughout surface with Cu30/W70 than that of Cu20/W80. These factors contribute to difference in temperature gradient of the workpiece which lead to difference in surface roughness

Table 10. Result of confirmation Experiment for Ra.

Response value	Best Conditions		Conditions	Difference
	Estimation	Experiment	Experiment	
Level	$A_2B_1C_1D_1$		$A_1B_1C_1D_1$	
Ra	2.47	2.68	2.70	0.05
S/N Ratio	-7.55		-8.01	

4. Conclusions

EDM on Ti6Al4V by two grade of copper tungsten using Taguchi's method is proposed. Results were summarized as follows:

1. In case of copper tungsten electrode, Cu30/W70 is better than Cu20/W80 in all cases for MRR, EWR and surface roughness.
2. The greatest machining condition to obtain maximum MRR was found to be $A_2B_3C_1D_2$, with a current level of 30 A, on-time level of 50 μs, and duty factor level of 40%.
3. To minimize EWR, the optimal machining parameters were identified as $A_2B_1C_3D_3$, with a current level of 5 A, on-time level of 400 μs, and duty factor level of 80%.
4. For achieving minimum Ra, the optimal machining parameters were identified as a current level of 5 A at level 1, on-time level of 50 μs at level 1, and duty factor level of 5% at level 1.
5. The confirmation experiment verified that the proposed approach is capable of identifying optimal process parameters.

5. References

[1] Ho K H and Newman S T 2003 *Int. J. Machine Tools and Manufacture* **43** p 1287
[2] Lin Y C Yan B H and Chang Y S 2000 *J.Materials Processing Technology* **104(3)** p 171

[3] Mahajan R Krishna H Singh A K and Ghadai R K 2018 India Int. Conf. Mechanical, Materials and Renewable Energy **377** p 1
[4] Muttamara A 2015 *Int. J. Chemical Engineering and Applications* **6**(**4**) p 251
[5] Wang Y L Liang S H Xiao P Yang X H and Zou J 2012 *Rare Metal Materials and Engineering* **41** p 266
[6] Yang X H Zou J T Xiao P and Wang X 2014 *Vacuum* **106** p 16
[7] Ranjit K 2001 *Design of Experiments Using the Taguchi Approach* (John Wiley & Sons)
[8] Peace G S 1993 *Taguchi Methods* (Addison-Wesley Publishing)
[9] Muttamara A and Phimoolchat J 2018 *Int J Mech Eng Technol,* **9(13)** p 978
[10] Lin J L Wang K S Yan B H and Tang Y S 2000 *Int. J. Adv. Manu. Technol.* **16** p 113
[11] Liu K Lauwers B and Reynaerts D 2010 *Int. J. Adv. Manu. Technol.* **47**(**1**) p 11

Acknowledgements
The authors acknowledge the support from Faculty of Engineering, Thammasat School of Engineering (TSE), Thammasat University.

Synthesis of Nano Porous Carbon derived from Vinasses Wastes for Crystal Violet Adsorption Application

Phetcharat Nenyoo[1], Anuchit Jaruvanawat[2], Atthapon Srifa[3], Worapon Kiatkittipong[4], Apiluck Eiad-Ua[2] and Suttichai Assabumrungrat[1]

[1] Faculty of Engineering, Chulalongkorn University, Bangkok 10330, Thailand

[2] College of Materials Innovation and Technology, King Mongkut's Institute of Technology, Ladkrabang, Bangkok 10520, Thailand

[3] Faculty of Engineering, Mahidol University, Nakhon Pathom 73170, Thailand

[4] Faculty of Engineering and Industrial Technology, Silpakorn University, Nakhon Pathom 73000, Thailand

E-mail: apiluck.ei@kmitl.ac.th

Abstract. An approach to removing crystal violet (CV) dye from aqueous solutions was investigated by nano porous carbon (NC) derived from vinasse waste. The NC was prepared via pyrolysis carbonization and potassium hydroxide chemical activation process. The NC was characterized by TGA, FTIR, and BET analysis. The optimal condition of the NC was used to adsorb CV dye in aqueous solutions at room temperature. Different initial CV concentrations and contact times were investigated to find the effect of these parameters. The NC was found efficient and can be used as a feasible alternative for removing CV dye from aqueous solutions. The NC was synthesized at activation temperature of 900 °C and the 1:1 precursor - activator (KOH) ratio provided the pore size of 3.285 nm and the maximum surface area 1,018 m^2/g, which was an optimal condition for the crystal violet adsorption. Crystal violet was highly adsorbed within 5 minutes and then was equilibrium within 30 minutes of the contact time. The amount of dye ion adsorbed onto adsorbent was 3.3, 7.2, 11.2, and 19.1 mg/g, respectively.

1. Introduction

Nowadays, the large-scale industries have expanded and increased. The textile industry is another industry that has a high extension rate, especially natural textile. It is a chemical industry that uses a lot of chemical dyes and water in the production process. The dye bleaching process releases wastewater that is contaminated with chemical dyes [1]. The chemical dyes affect the organisms that live in the water. It also blocks sunlight from entering the water, reducing the rate of photosynthesis, and affecting the food chain [2]. Cationic dyes are more harmful to living organisms than anionic dyes due to the cationic dye can interact with anion at the surface of the cell membrane [3]. Although the dye-contaminated wastewater is low toxicity, it is a dark pigment and difficult to decompose [4] due to its light and pH resistance [5]. Therefore, it is essential to eliminate chemical dyes before being released into the water source.

Crystal violet (CV) or Gentian violet is a cationic dye. Crystal violet is the most widely used in pharmaceutical production, classification of bacteria as a staining agent, and purple colouring textile products. In addition, it is also used in poultry feed to prevent fungal growth due to its antimicrobial activity [2]. However, crystal violet is classified as a hazardous substance because it has a triphenylmethane group. Its carcinogenic and mutagenic nature. It can affect the cell division process and cause lasting damage to the eye, especially the conjunctiva and cornea. Moreover, the release of the dyes harms the environment [6]. For this reason, it is essential to prevent the appearance of crystal violet in the environment by using the appropriate methods.

There are several methods for wastewater treatment in the dyeing industry, such as biological process, flocculation, membrane separation, coagulation, bioaccumulation, and electrochemical methods [6,7]. Each method has different limitations. Currently, there are photocatalysis processes and adsorption with magnetic particles for use in dye treatment. Although such methods have high efficiency, it has a high cost and is not widely applicable in the industry [7]. Therefore, the adsorption process is another favourite method due to it is low cost, high efficiency, and not complicated. The adsorption operation can be carried out at room temperature [4]. The most widely used adsorbent in the industry is activated carbon due to its high surface area and the ability to synthesize a pore structure [5]. However, the limitation of activated carbon is the high production cost, and the reuse of the adsorbent has lower adsorption efficiency [1].

In this study, the crystal violet removal problem is presented by using nano porous carbon derived from vinasse waste as an absorbent. The vinasse waste is a large by-product of the sugar or ethanol industry. After the removal of the desired product (alcohol, ascorbic acid, etc.), the remaining material is called vinasse. Vinasse has a dark brown liquid and unpleasant odour. Moreover, it has a high concentration of acid and organic matter. In general, vinasse is suitable for raising all kinds of animals. It's used as a substitute for molasses. Yeast and various nutrients resulting from complete fermentation make vinasse have a high nutritional value. It's also used as a bio-organic fertilizer. However, vinasse is acidic, thus affecting soil and water quality. Therefore, it is essential to dispose of this waste to prevent any environmental impact [8].

In the present study, nano porous carbon was synthesized from vinasse waste by pyrolysis carbonization and chemical activation process with potassium hydroxide (KOH). To study the effect of activation temperature and KOH concentration that affect the formation of pore structure. In the adsorption process, nano porous carbon has been used. The ability and efficiency of nano porous carbon in removing crystal violet dye from an aqueous solution were evaluated. In addition, the effects of parameters such as initial crystal violet dye concentration and contact time on the crystal violet dye adsorption process were investigated.

2. Materials and Method

2.1 Materials

Vinasse waste was used as the source of the carbon, collected from Mitr Phol sugar factory located at Chaiyaphum, Thailand. Vinasse was dried at 150°C for 48 hours in the oven, then grinded and sieved into a vinasse powder. Commercial crystal violet from Vidhyasom Co., Ltd. located at Bangkok, Thailand. Potassium hydroxide (KOH) laboratory-grade were used in this study.

2.2 Synthesis of nano porous carbon

The nano porous carbon (NC) was prepared via pyrolysis carbonization and chemical activation by using KOH as an activating agent. First, the vinasse powder was pyrolyzed in furnace tube at 800 °C for 1 hour under the nitrogen atmosphere. After that, mixing the carbon powder with 1, 2, 3 of KOH concentration by a mass ratio for 15 min. The mixture was activated in furnace tube at 600, 700, 800, and 900°C for 1 hour under the nitrogen atmosphere. Finally, the residue on nano porous carbon was washed with deionized water and dried in the oven at 105°C for 24 hours. The physical and chemical properties of nano porous carbon was characterized by thermogravimetric analysis (TGA), infrared spectroscopy (FTIR), and Nitrogen sorption isotherm.

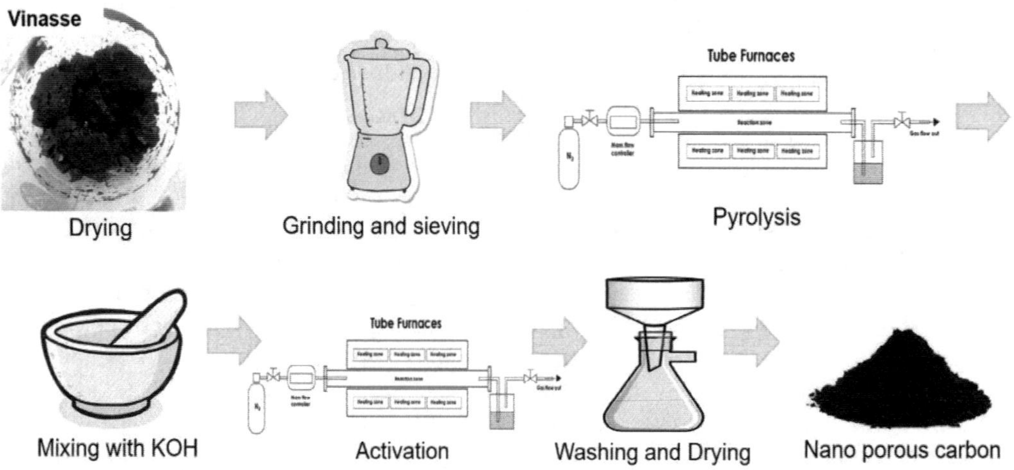

Figure 1. synthesis of nano porous carbon process.

2.3 Crystal violet adsorption

The optimal condition of nano porous carbon was used as adsorbent material. In adsorption process at room temperature, initial crystal violet concentration of 1, 2, 3, and 5 mg/L, 50 mg/L of adsorbent dose and contact time from 0 to 60 minutes was investigated. By using a spectrophotometer, the dye ions were measured at their initial and equilibrium concentrations. The amount of dye ion adsorbed onto adsorbent is calculated using equation (1):

$$q_t = \frac{C_0 - C_t}{W} \times V \quad (1)$$

To calculate the adsorption efficiency (%Adsorption) of the dye ion, equation (2) can be used as:

$$\%\,Adsorption = \frac{C_0 - C_t}{C_0} \times 100 \quad (2)$$

Here q_t shows the amount of dye ion taken up per unit mass (mg/g), V is the volume of adsorbate solution in liter, W is mass (g) of dry adsorbent, C_0 and C_t are the initial and equilibrium concentrations (mg/L) of adsorbate, respectively.

3. Results and Discussions

3.1. Thermal gravimetric analysis

The thermal behaviour of the vinasse powder is shown in the figure 2. The mass loss profile as a function of temperature within the temperature range of 25°C to 900°C. The overall mass loss of vinasse powder can be divided into five phases. The first phase corresponds to the removal of free moisture, bond moisture, and evaporation of volatile matter in the range of 25°C to 100°C and 100°C to 200°C, respectively. In the second phase, there occurs the main volatilization and a mass loss of ~30% is achieved. Then a mass loss of ~5%, followed by a weight loss of 10% with a decomposition. It can be seen in the range of 715°C to 795°C. Finally, no mass loss was detected up to 900°C. The vinasse was decomposed at high temperatures.

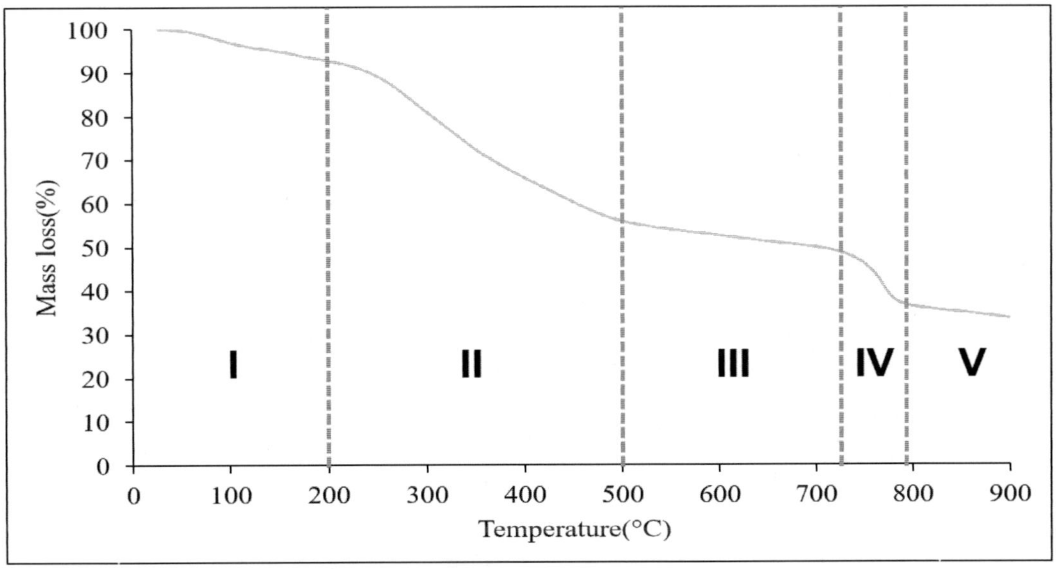

Figure 2. Thermal gravimetric analysis (TGA) of the vinasse powder, temperature range of 25°C to 900°C, and under nitrogen atmosphere.

3.2. Fourier-transform infrared spectroscopy (FTIR)

The wavenumber position and relative intensity of the FTIR spectra were used to evaluate the transformation that the vinasse underwent in the pyrolysis carbonization under different conditions. Therefore, the significant change in the FTIR spectra profile of vinasse powder were observed that compare to the vinasse pyrolyzed at 800°C for 1 hour.

The FTIR spectra of the vinasse powder is shown in the figure 3a. In general, the wide bands at 3500 cm-1 to 3000 cm-1 of the vinasse powder can be attributed to the O-H stretches of adsorbed water molecules, alcohols, and carboxylic acids. The C-H stretches can be observed in the 2990 cm-1 to 2800 cm-1 region. At 1610 cm-1 bands related to C=C vibrations of aromatic rings. In addition, 1373 cm-1 bands attributed to the deformation of aliphatic C-H. Approximately, 1105 cm-1 for the C-O of alcohols, phenols, and vibrations of the remaining glycosidic skeleton from cellulose suspend in the vinasse. At 600 cm-1 bands related to the C-H bending.

In the figure 3b. is shown the FTIR spectra of the vinasse pyrolyzed at 800°C for 1 hour. The intensity decreases of the band in the vinasse pyrolyzed when compared to that observed in the vinasse powder. The intensity reduction of the first band suggested that hydrolysis and dehydration occurred during the pyrolysis carbonization of the vinasse powder.

The FTIR spectra of nano porous carbon is shown in the figure 4. The intensity decreases of the bands related to the C-H deformation, C-O alkoxy, and C-H bending when increased the activation temperature from 600°C to 900°C.

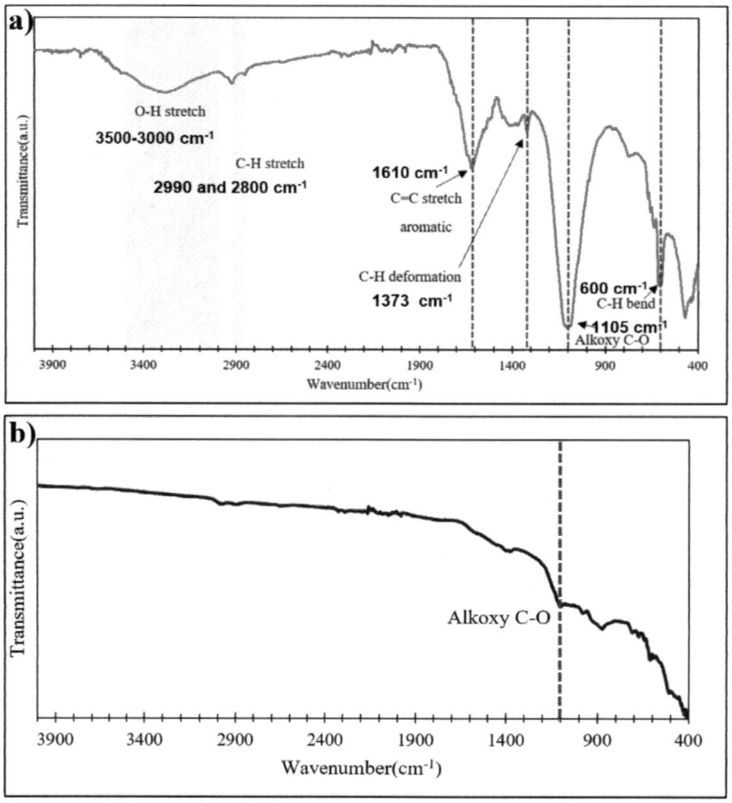

Figure 3. FTIR spectra of (a) vinasses powder and (b) vinasse pyrolyzed at 800°C for 1 hour.

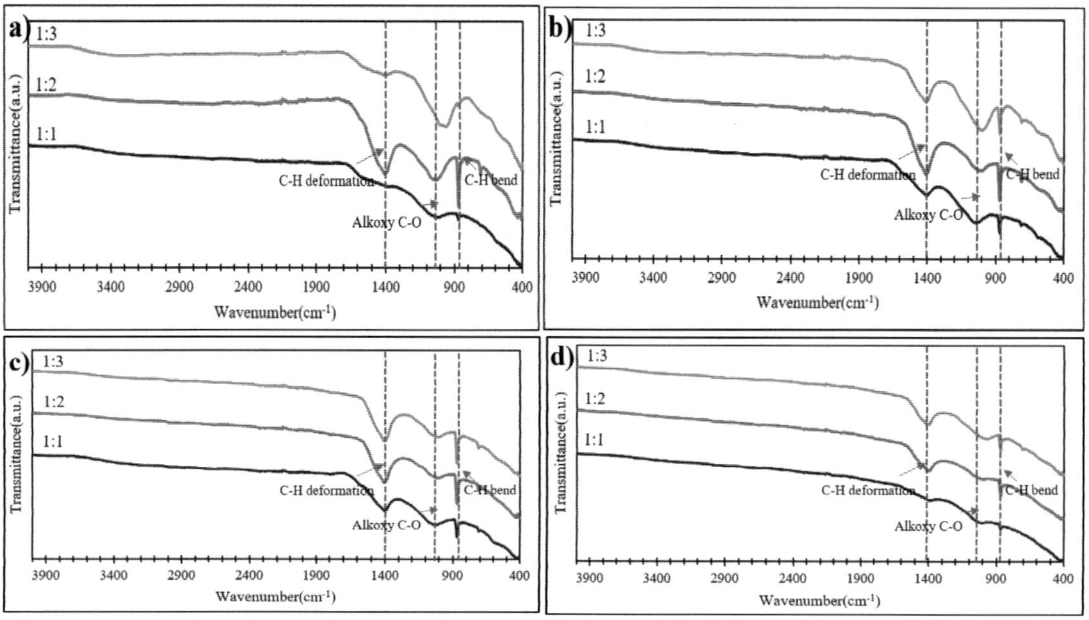

Figure 4. FTIR spectra of nano porous carbon activated with KOH from 1:1, 1:2, and 1:3 by a mass ratio of carbon: KOH at (a) 600°C, (b) 700°C, (c) 800°C, and (d) 900°C at 1 hour under nitrogen flow.

3.3. *BET analysis of nitrogen adsorption-desorption isotherm*

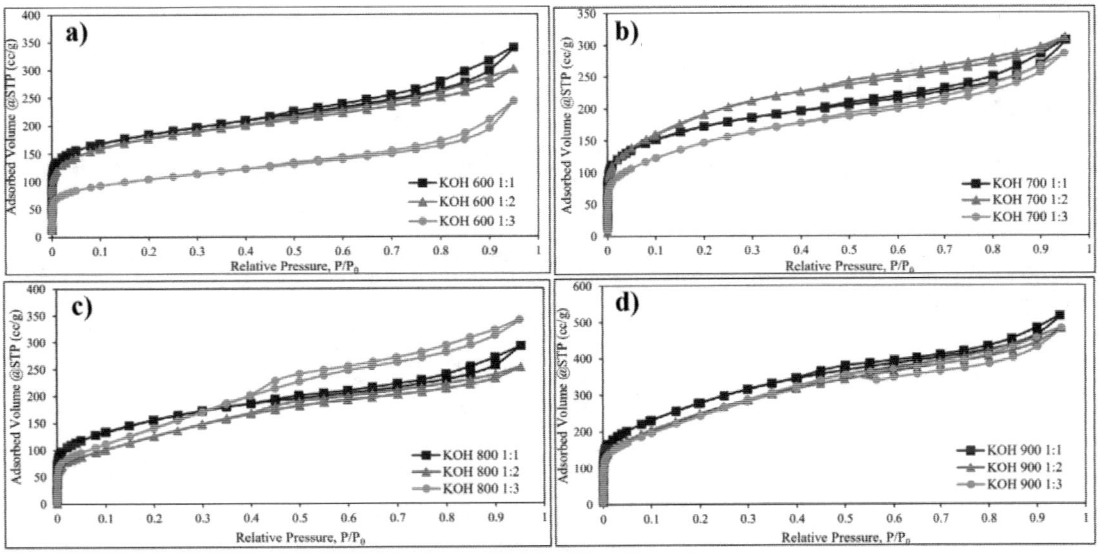

Figure 5. N_2 sorption isotherm of nano porous carbon activated with KOH from 1:1, 1:2, and 1:3 by a mass ratio of carbon: KOH at (a) 600°C, (b) 700°C, (c) 800°C, and (d) 900°C at 1 hour under nitrogen flow.

The nitrogen adsorption-desorption isotherms of nano porous carbon are shown in the figure 5. The shape of N_2 sorption isotherms were different at different activation temperature and KOH concentration. These isotherms belong to a type IV of IUPAC classification. A type IV isotherm indicated a mixture of microporous and mesoporous structure.

The summary of BET surface area, total pore volume, and average pore diameter of nano porous carbon at different conditions is shown in the table 1. The surface area was decreased when increased the KOH concentration from 1:1 to 1:3 by a mass ratio. At activation temperature of 900°C and 1:1 of precursor-KOH ratio was the optimal condition of nano porous carbon that provided the maximum surface area of 1,018 m^2/g, the total pore volume of 0.836 cc/g, and average pore diameter of 3.285 nm. The optimal condition of nano porous carbon was used as adsorbent material.

IOP Conf. Series: Materials Science and Engineering 1286 (2023) 012002 doi:10.1088/1757-899X/1286/1/012002

Table 1. Comparison of the basic physical properties for nano porous carbon.

	Concentration Precursor: KOH	S_{BET} (m^2/g)	Total pore volume (cc/g)	Average pore diameter (nm)
Bare 800°C 1 h	-	72.02	0.074	4.144
NC 600°C 1 h	1:1	671.60	0.525	3.130
	1:2	640.80	0.466	2.907
	1:3	375.00	0.377	4.018
NC 700°C 1 h	1:1	620.50	0.475	3.061
	1:2	691.10	0.482	2.788
	1:3	529.00	0.442	3.342
NC 800°C 1 h	1:1	558.50	0.451	3.233
	1:2	469.00	0.391	3.337
	1:3	543.60	0.528	3.882
NC 900°C 1 h	1:1	1018.00	0.836	3.285
	1:2	910.40	0.745	3.275
	1:3	915.4	0.746	3.259

3.4. Crystal violet adsorption

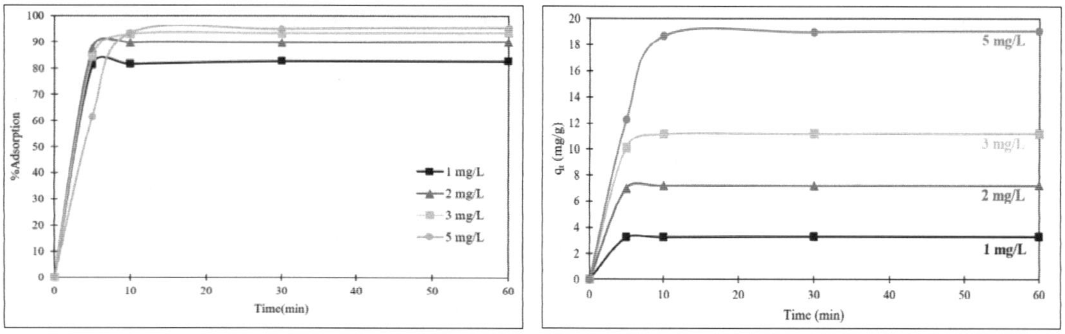

Figure 6. Contact time effect of crystal violet adsorption at room temperature, adsorbent dose 0.05 g/L, and initial crystal violet concentration of 1, 2, 3, and 5 mg/L, respectively. (a) the adsorption efficiency and (b) the amount of dye ion adsorbed onto adsorbent.

The adsorption efficiency of crystal violet adsorption and contact time from 0 to 60 minutes is shown in the figure 6a. The initial crystal violet concentration of 1, 2, 3, and 5 mg/L at 5 minutes of the contact time obtained the adsorption efficiency of crystal violet adsorption up to 81, 87, 84, and 64%, respectively. Then slightly increased until 30 minutes was equilibrium. From the figure 6b. the amount of dye ion adsorbed onto adsorbent was 3.3, 7.2, 11.2, and 19.1 mg/g, respectively.

4. Conclusion

Vinasse wastes can be used as a raw material for synthesis of nano porous carbon. The nano porous carbon was synthesized at activation temperature of 900 °C and the 1:1 precursor - activator (KOH) ratio provided the pore size of 3.285 nm and the maximum surface area 1,018 m^2/g, which was a suitable condition for the crystal violet adsorption. Crystal violet was highly adsorbed within 5 minutes and then was equilibrium within 30 minutes of the contact time. The amount of dye ion adsorbed onto adsorbent was 3.3, 7.2, 11.2, and 19.1 mg/g, respectively.

5. Acknowledgment

This work has partially been supported by Faculty of Engineering, Chulalongkorn University, College of Materials Innovation and Technology, King Mongkut's Institute of Technology, Faculty of Engineering, Mahidol University, and Faculty of Engineering and Industrial Technology, Silpakorn University.

6. Reference

[1] Shakoor, S.; Nasar, A. Adsorptive Decontamination of Synthetic Wastewater Containing Crystal Violet Dye by Employing Terminalia Arjuna Sawdust Waste. *Groundw. Sustain. Dev.* 2018, 7, 30-38.

[2] Chakraborty, S.; Chowdhury, S.; Saha, P. Das. Insight into Biosorption Equilibrium, Kinetic and Thermodynamics of Crystal Violet onto *Ananas Comosus* (Pineapple) Leaf Powder. *Appl. Water Sci.* 2012, 2(2), 135-141.

[3] Miyah, Y.; Lahrichi, A.; Idrissi, M.; Boujraf, S.; Taouda, H.; Zerrouq, F. Assessment of Adsorption Kinetics for Removal Potential of Crystal Violet Dye from Aqueous Solutions Using Moroccan Pyrophyllite. *J. Assoc. Arab Univ. Basic Appl. Sci.* 2017, 23, 20-28.

[4] Paven, F. A.; Camacho, E. S.; Lima, E. C.; Dotto, G. L.; Branco, V. T. A.; Dias, S. L. P. Formosa Papaya Seed Powder (FPSP): Preparation, Characterization and Application as an Alternative Adsorbent for the Removal of Crystal Violet from Aqueous Phase. *J. Environ. Chem. Eng.* 2014, 2(1), 230-238.

[5] Rawinipa, S. Wastewater Dye Treatment via Adsorption Process. Science Journal. Khonkan University. 2016 44(3): 419-434.

[6] Saeed, A.; Sharif, M.; Iqbal, M. Application Potential of Grapefruit Peel as Dye Sorbent: Kinetics, Equilibrium and Mechanism of Crystal Violet Adsorption. *J. Hazard. Mater.* 2010, 179(1-3), 564-572.

[7] Silverira, M. B.; Paven, F. A.; Gelos, N. F.; Lima, E. C.; Dias, S. L. P. Punica Granatum Shell Preparation, Characterization, and Use for Crystal Violet Removal from Aqueous Solution. *Clean-Soil, Air, Water* 2014, 42(7), 939-946.

[8] Hidalgo, K. Vinasse in Feed: Good for Animal and Environment. *Feed Tech.* 2019. 13(5):18-20.

IOP Conf. Series: Materials Science and Engineering 1286 (2023) 012003 doi:10.1088/1757-899X/1286/1/012003

Synthesis of Lignin Extracted from Black Liquor to Nano Porous Carbon Materials for Tetracycline Adsorption Application

Tassanai Tempiam[1], Anuchit Jaruvanawat[2], Atthapon Srifa[3], Worapon Kiatkittipong[4], Apiluck Eiad-Ua[2,*] and Suttichai Assabumrungrat[1]

[1]Faculty of Engineering, Chulalongkorn University, Bangkok 10330, Thailand

[2]College of Materials Innovation and Technology, King Mongkut's Institute of Technology, Ladkrabang, Bangkok 10520, Thailand

[3]Faculty of Engineering, Mahidol University, Nakhon Pathom 73170, Thailand

[4]Faculty of Engineering and Industrial Technology, Silpakorn University, Nakhon Pathom 73000, Thailand

E-mail: apiluck.ei@kmitl.ac.th

Abstract. Tetracycline is an antibiotic that suppresses the growth of bacteria, treats infections in the urinary tract, treats intestinal inflammation, bronchitis, treats dysentery, treats wounds, abscesses, pus, various inflammations due to infections. However, antibiotics have attracted extensive attention from the scientists because a variety of antibiotics have been. detected in soil, surface water, groundwater, and even in drinking water in the recent years. Antibiotics are among the most potent pollutants due to their overuse. This is because most antibiotics are metabolized and poorly absorbed by humans and treated animals, so many fractions are excreted through urine and feces into water. Therefore, it is urgent to remove tetracycline from water environment. There are several techniques used to remove tetracycline from aqueous solutions, coagulation, photo electrocatalytic degradation, ion exchange, membrane processing, and adsorption. Among them, adsorption is a widely used efficient technology, because it has the advantages of easy operation, low cost, high efficiency, and there is no risk of highly toxic by-products. In this work, lignin from black liquor in chemical pulping processes is used as a precursor for nano porous carbon synthesis. The physical and chemical properties of nano porous carbon was investigated by field-emission scanning electron microscope (FESEM), X-ray diffractometer (XRD), Energy Dispersive X-ray Spectroscopy (EDX), and Nitrogen sorption isotherm, respectively. The maximum surface area appears at 1,676.5 m^2/g and a total pore volume of 1.091 cm^3/g, this sample was used as an adsorbent and could perform its functions perfectly.

1. Introduction

Tetracycline (TC) is an antibiotic that inhibits the growth of bacteria, treat urinary tract infections, treat intestinal inflammation, bronchitis, treat dysentery, treat wounds, abscesses, pus, various inflammations due to infection. However, antibiotics have received widespread attention from scientists. This is because many antibiotics have been detected in soil, surface water, groundwater and even drinking water in recent years. Antibiotics are one of the most potent pollutants due to their overuse. This is because most antibiotics are metabolized and poorly absorbed by humans and treated animals. So many fractions are excreted in the urine and feces into the water [1]. Therefore, it is urgent to remove tetracycline from water environment. There are several techniques used to remove tetracycline from aqueous solutions, coagulation, photo electrocatalytic degradation, ion exchange, membrane processing, and adsorption. Among them, adsorption is a widely used efficient technology, because it has the advantages of easy operation, low cost, high efficiency, and there is no risk of highly toxic by-products [2]. In this work, lignin from black liquor (BL) in chemical pulping processes is used as a precursor for nano porous carbon synthesis. Because black water is a waste in the pulp industry, which consist mainly of solid lignin and lignin with high carbon composition suitable for use as a precursor for nano porous carbon synthesis [3]. The purpose of this work is to synthesis waste Black liquor to nano porous carbon and used to tetracycline adsorption.

2. Materials and Methods

2.1. BL precipitation

BL waste courtesy of SIAM KRAFT Industry Co., Ltd., Wangsala Plant, SCG Packaging. In this work we used a commercially available Potassium alum (Alum) for BL precipitation. BL pH12 (1000mL) was adjust to pH4 by adding Alum until pH4, then precipitate overnight. The elemental analysis was investigated by), Energy Dispersive X-ray Spectroscopy (EDX). Material crystalinity was verified by X-ray diffractometer (XRD).

2.2. Synthesis of nano porous carbon

The porous carbon (PC) was prepared by chemical activation (carbonization) technique by using KOH activating agent. Figure 1. Shows the precipitates BL was carbonize at various temperature 600 – 900° C under nitrogen flow by using KOH activating at 50%wt – 150%wt for 2hrs. The condition was called PCxxx-y.y (x is temperature and y is activating agent to precursor ratio).Surface morphology was investigated by Field-emission scanning electron microscope (FESEM). Surface area and pore characteristic was analyzed by Nitrogen sorption isotherm.

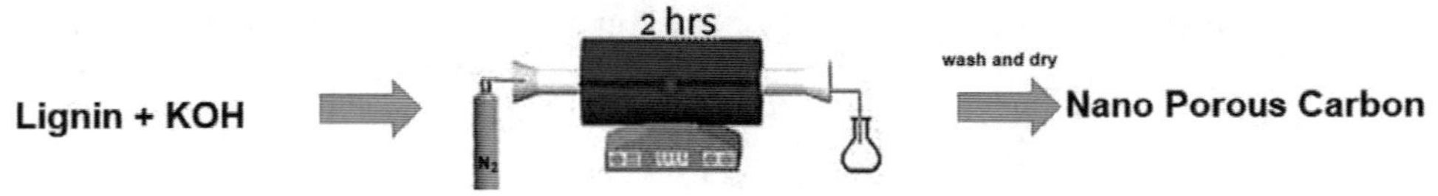

Figure 1. *nano porous synthesized diagram*

2.3. TC adsorption

The optimal condition of nano porous carbon (maximum surface area) was used as adsorbent material. In experiments, TC was used at a concentration of 5-20ppm and 500mg/L of adsorbent, adsorption time from 0 -180 minutes.

3. Results and Discussions

3.1. BL and precipitated BL comparison

The results of the elemental composition analysis showed that the composition of the reactants was consistent with other studies [4]. Table 1 shows the high composition of elements Na and K due to the main product of BL waste was produced by the alkaline process. The precipitation process shows a clear change in which the content of elements Na and K is greatly reduced Table 2. This precipitation process results in a nearly 50% higher carbon content of the precursors, which is suitable for synthesis into nano porous carbon materials

Table 1. Element composition of BL

Element	Weight%	Atomic%
C	44.22	54.15
O	39.65	36.46
Na	12.22	7.82
K	2.56	0.96
Other	1.36	0.61

Note The substance has been dried.

Table 2. Element composition of precipitated BL

Element	Weight%	Atomic%
C	60.43	67.42
O	37.89	31.74
Na	0.07	0.04
K	0.03	0.02
Other	1.58	0.77

Figure 2 shows a clear contaminant of BL. The precipitation process XRD clearly shows that the contaminants can be removed. The precipitated BL peak is characterized by a broad peak, the appearance matches the lignin XRD curve [5]. The results of elemental and structural analysis show that the counterfeit is removed through the precipitation process.

IOP Conf. Series: Materials Science and Engineering 1286 (2023) 012003 doi:10.1088/1757-899X/1286/1/012003

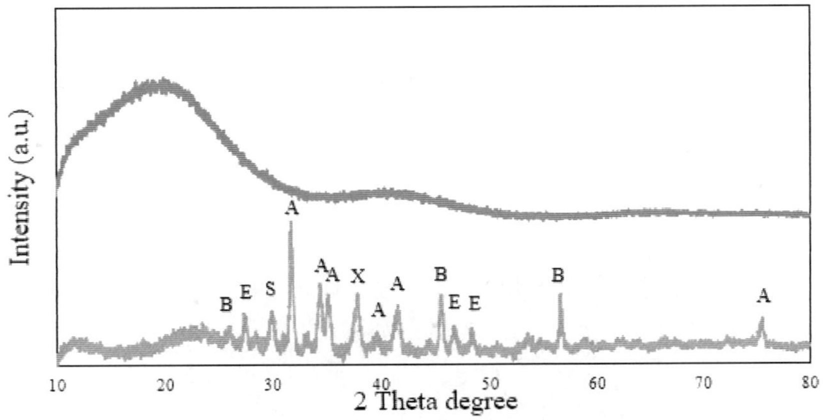

Figure 2. *XRD Pattern of precipitated BL (top) and dried BL (bottom) Phases were identified as: A–orthorhombic sodium carbonate; S–monoclinic sodium sulfate; E–monoclinic potassium carbonate; X–cubic calcium oxide and B–orthorhombic calcium carbonate*

3.2. Nano porous carbon characteristic

Figure 3 shows the clear pore structure of product. Figure 3. a) PC700-1.0, honeycomb-like porosity can be observed. An increase in the synthesis temperature causes the pore structure to begin to collapse at temperatures of 700oC and above. This is because too much temperature increases can cause partial carbon decomposition, causing the structure to begin to collapse as the temperature increases above 700 °C [6]. From Table 3 it is clearly shown that the activation process can increase the surface area of the material. The resulting pore is at the meso-pore level considering the diameter. The Activation temperature at 700 °C and the 1:1 activating agent (KOH)- precursor ratio provided the maximum surface area of 1,676.5 m^2/g and a total pore volume of 1.091 cm^3/g. Table 3 corresponds to the FESEM image demonstrating PC700-1.0, which exhibits clear pore characteristics and maximum surface area. The proper %wt. activating agent ratio is 1:1, since at all temperatures this activating agent ratio provides the highest surface area.

Table 3. Pore characteristic of nano porous carbon

Condition	Average D (nm)	Pore characteristics	
		SBET (m^2/g)	V_T (cm^3/g)
PC600-0.0	5.4	44.0	0.060
PC600-0.5	2.6	727.8	0.476
PC600-1.0	2.5	792.3	0.498
PC600-1.5	2.5	787.4	0.484
PC700-0.0	5.4	41.9	0.057
PC700-0.5	2.7	1051.0	0.701
PC700-1.0	2.6	1676.5	1.091
PC700-1.5	2.4	927.3	0.561
PC800-0.0	4.4	52.1	0.058
PC800-0.5	2.8	965.1	0.641
PC800-1.0	2.6	1346.9	0.944

PC800-1.5	2.4	1307.2	0.782
PC900-0.0	4.2	56.2	0.059
PC900-0.5	2.6	978.6	0.915
PC900-1.0	3.4	1187.8	1.009
PC900-1.5	5.3	936.0	1.241

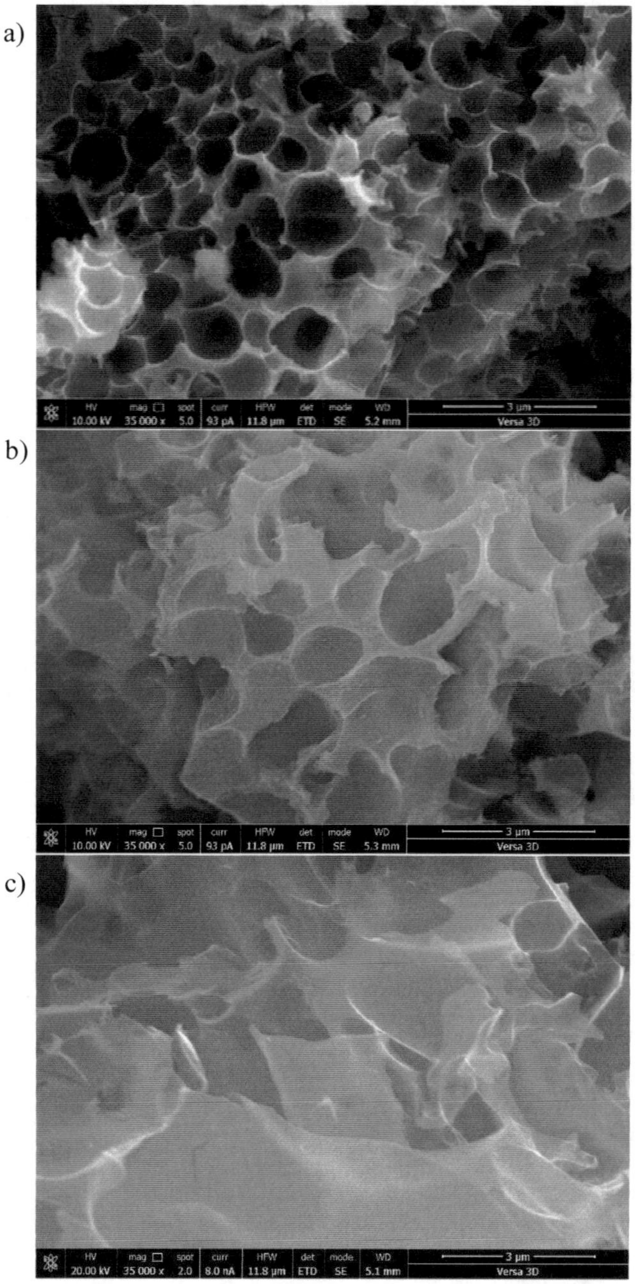

Figure 3. *FESEM image of a) PC700-1.0 b) PC800-1.0 c) PC900-1.0*

The mechanism of activation process, KOH was converted to K_2CO_3 via the reactions between KOH and O-containing carbon and transformed to K_2O. These processes promoted porosity of carbon. [8]

3.3. TC adsorption

From figure 4 an increase in TC lead to increase in adsorption percentage and adsorption capacity. Absorption times of since 120 demonstrated constant adsorption capacity. Increasing the amount of adsorbent tends to increase the TC adsorption percentage because increasing the adsorbent content increases the surface area and adsorption site [7].

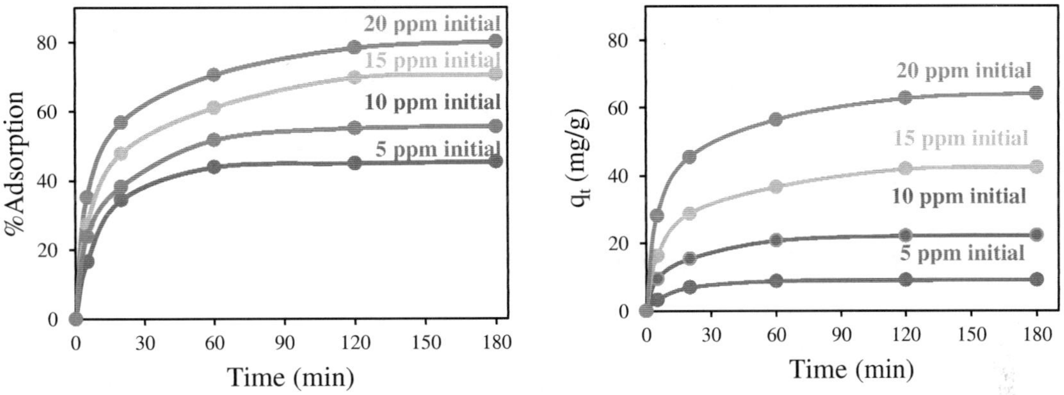

Figure 4. *(a) % Adsorption and (b) Adsorption capacity*

4. Conclusion

The environmentally friendly precipitating process for Black liquor to Lignin purification was investigated in lab-scale experiments. Black liquor from pulping process demonstrated to be a suitable precursor carbonization process, to prepare the nano porous carbon. Finally, the industrial waste black liquor could be converted to nano porous carbon and utilized to adsorption of TC antibiotic from water solution.

Acknowledgments

This work was financially supported by the Faculty of Engineering Chulalongkorn University, College of Materials Innovation and Technology, King Mongkut's Institute of Technology, Ladkrabang.

References

[1] Daghrir, Rimeh et Drogui, Patrick (2013). Tetracycline antibiotics in the environment: a review. Environmental Chemistry Letters, vol. 11, n° 3. p. 209-227.

[2] Lenka BLINOVÁ, Maroš SIROTIAK Research Papers Faculty of Materials Science and Technology Slovak University of Technology, Volume 29, Issue 48, pp.22-36

[3] Jun'ichiHayashi, AtsuoKazehayaa, Katsuhiko Muroyamaa, A.Paul Watkinsonb. Carbon Volume 38, Issue 13, 2000, Pages 1873-1878

[4] Esa KariVakkilainen Steam Generation from Biomass Construction and Design of Large Boilers 2017, Pages 237-259

IOP Conf. Series: Materials Science and Engineering 1286 (2023) 012003 doi:10.1088/1757-899X/1286/1/012003

[5] Abla Alzagameem, Basma El Khaldi-Hansen, Dominik Büchner, Michael Larkins, Birgit Kamm, Steffen Witzleben, Margit Schulze Molecules. 2018 Oct 16;23(10):2664

[6] Wan Mohd Ashri WanDaudWan Shabuddin WanAliMohd ZakiSulaiman Carbon Volume 38, Issue 14, 2000, Pages 1925-1932

[7] Ho, Y.S. and Ofomaja, A.E. 2006. Biosorption thermodynamic of cadmium on coconut copra meal as biosorbent. Biochemical Engineering Journal. 30: 117-123.

[8] Wei Chen, Meng Gong, Kaixu Li a, Mingwei Xia, Zhiqun Chen, Haoyu Xiao, Yang Fang, Yingquan Chen, Haiping Yang and Hanping Chen, Insight into KOH activation mechanism during biomass pyrolysis: Chemical reactions between O-containing groups and KOH, Applied Energy Volume 278, 15 November 2020,

Preparation of Titanium Dioxide Nanoparticles Films by Air-Compressed Spray Method for Supercapacitor Device

Benchapol Tunhoo[1], Kanyawee Apiruktatarn[1], Korakot Onlaor[1,*] and Thutiyaporn Thiwawong[1]

[1] Electronics and Control System for Nanodevices Research Laboratory, College of Materials Innovation and Technology, King Mongkut's Institute of Technology Ladkrabang, Bangkok, 10520, Thailand

*E-mail: korakot.onlaor@gmail.com

Abstract. In this work, the titanium dioxide (TiO_2) nanoparticle films were prepared from commercial Degussa P25-TiO_2 nanoparticles by air-compressed spray. The properties of the prepared films were characterized by X-ray diffraction, FTIR, Raman spectroscopy, and scanning electron microscope, respectively. In addition, the prepared TiO_2 film on a metal substrate (stainless steel, SS304) was used to fabricate the electrode of the electrochemical supercapacitor device. The electrochemical storage performance of the fabricated device was obtained by cyclic voltammetry measurement in aqueous 1 M Na_2SO_4 with an Ag/AgCl reference electrode (platinum as a counter electrode). The optimum specific capacitance of the fabricated device was found at 3.014 F/g at a scan rate of 50 mV/s.

1. Introduction

Metal oxide is a large group of materials widely used in many applications. TiO_2 is a category in the family of transition metal oxides [1]. Typically, the TiO_2 material exhibited many crystalline structures such as anatase, rutile, brookite, etc. TiO_2 can be used in various applications such as photocatalyst [2], solar cell [3], sensor [4] and, others. The Degussa P25-TiO_2 is a commercial grade of TiO_2 nanoparticles that was composed of separate phases between rutile (25%) and anatase (75%) structures. P25-TiO_2 is widely studied in many fields of applications due to its properties demonstrated high activity levels.

A supercapacitor is an energy storage device that can provide high energy and power density. One of the requirements of a supercapacitor device is an electrode, which collects the charge for storing energy in the device. It had been reported about various materials for fabrications of an electrode in supercapacitor device, including carbon-based, spinel metal oxide, etc. One of those is a TiO_2 material used as an electrode for a supercapacitor device [5].

In this work, the P25-TiO_2 nanoparticles films were prepared by air-compresses spray deposition technique on the metal substrate (stainless steel, SS304) at various deposition layers. The prepared films were characterized by X-ray diffractometer (XRD), Fourier-transform infrared spectroscopy (FTIR), and scanning electron microscope (SEM), respectively. In addition, the prepared TiO_2 films were applied as an electrode in a supercapacitor device and the properties of the devices based on P25-TiO_2 nanoparticles were studied and reported.

2. Experiment

In this work, the P25-TiO_2 nanoparticles were purchased from Aldrich Chem. Co, with an average particle size of 50 nm. The nanoparticles were dispersed in 1:1 by volume of deionized water and ethanol with a concentration of 0.5 mg/ml, which was used as a precursor solution. After that, the precursor solution was sonicated for 30 min to get a good dispersion solution. Then, the commercial air-compressed spray was applied to prepare the films. Each layer of films was prepared at a deposition time of 3 seconds with a substrate temperature of 70 °C. The deposition layers varied at 4, 6, 8, and 10. The physical properties of prepared TiO_2 films were performed by X-ray diffractometer (Rigaku, Smart Lab), Raman spectroscopy (Thermo Fisher, DXR SmartRaman), Fourier-transform infrared spectroscopy (PerkinElmer Scientific, Spectrum Two), and scanning electron microscope (Zeissm EVO HD), respectively. For the preparation of the supercapacitor electrode, the stainless steel (SS304) was used in the film area of 10×10 mm^2. A portable potentiostat (Dropsens, uStat 400) was used to investigate the electrochemical properties of prepared films.

3. Result

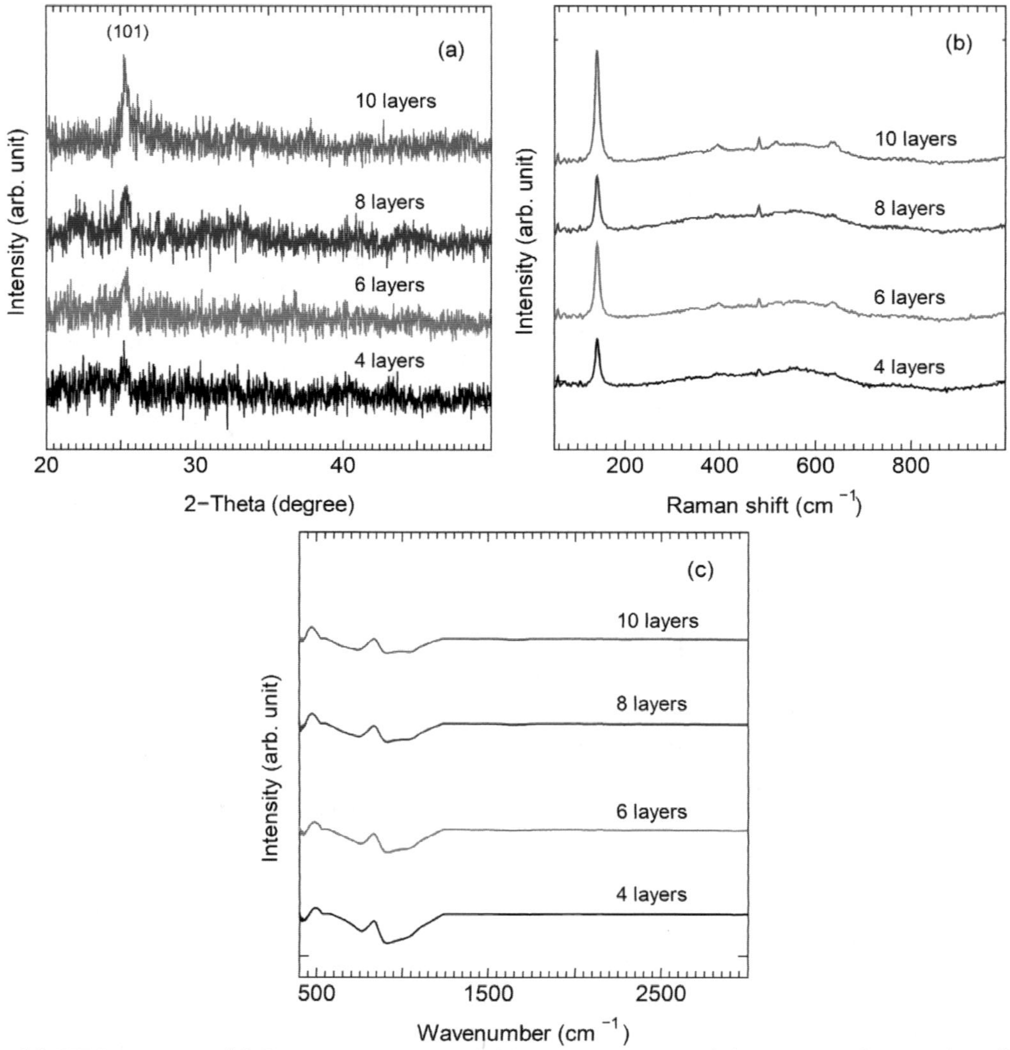

Fig. 1. (a) XRD spectra, (b) Raman spectra, and (c) FTIR spectra of the prepared P25-TiO_2 films at various deposition layers.

Fig.1(a) depicts XRD spectra of prepared TiO_2 films at various deposition layers from 2 to 10. The crystallinity of the prepared films depended on deposition layers. The diffraction peak of all deposition films can be observed at a diffraction angle of 25.4 degrees, that corresponding to a diffraction peak of the anatase phase. The XRD pattern of P25-TiO_2 showed the diffraction of 25.3 and 27.4 degrees which corresponded to the crystalline phase of anatase and rutile structure [6].

The Raman spectra of prepared TiO_2 films are shown in Fig.1(b). From the results, the Raman peaks located at 140, 395, 482, and 639 cm^{-1}, corresponding to Eg, B1g, A1g + B1g, and Eg of the Raman peaks of anatase phase TiO_2 [7].

FTIR spectra of prepared TiO_2 films are shown in Fig.1(c), which correspond with the characteristics of TiO_2 material. The FTIR peak at a wavenumber of 590 cm^{-1} was the vibration of Ti-O-O bonding [8]. It was not observed the peak at 1779 that was the characteristic of C=O. The FTIR spectra confirmed the presence of Ti-O bond in the prepared films.

Fig.2(a-d) depicts the surface morphologies of prepared TiO_2 films. The P25-TiO_2 nanoparticles can be seen on the top of the surface.

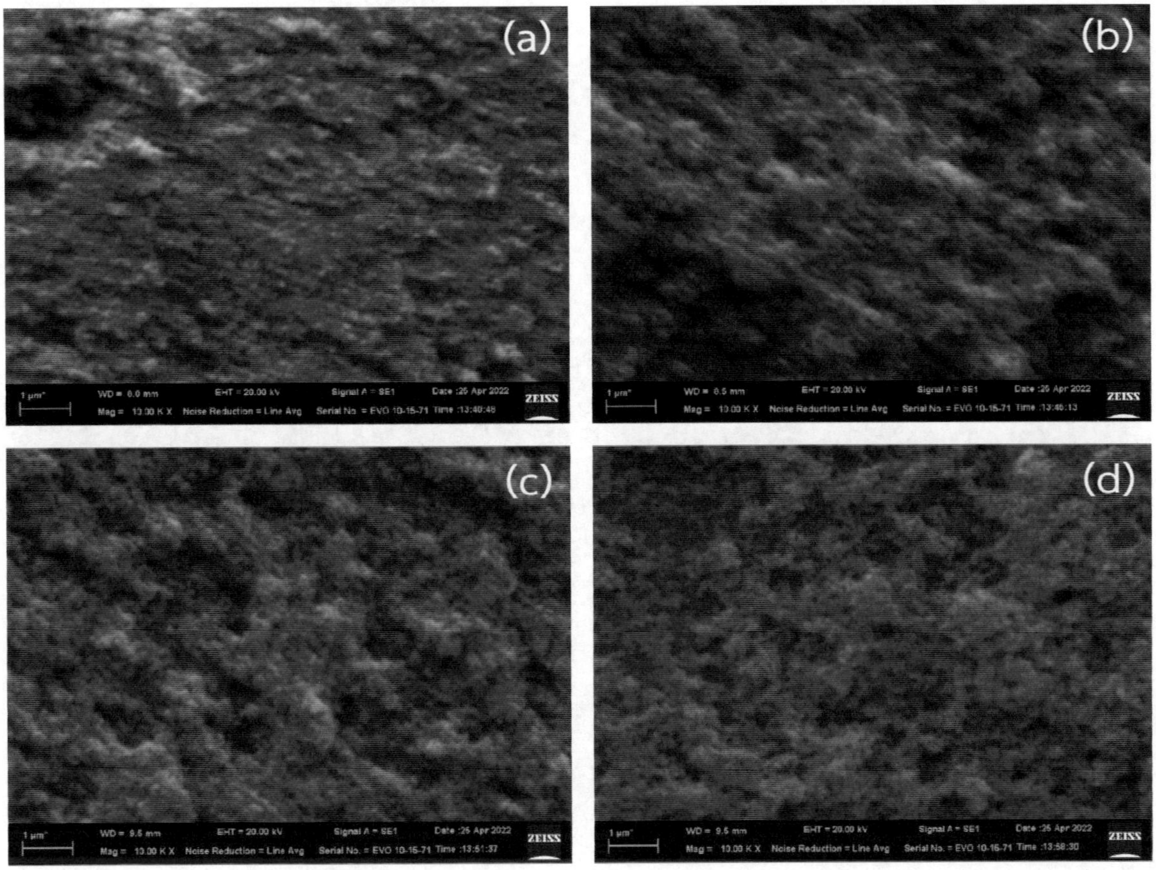

Fig. 2. SEM images of the prepared P25-TiO_2 films with different deposition layers (a) 4 layers, (b) 6 layers, (c) 8 layers, (d) 10 layers

The electrochemical measurement of the prepared TiO_2 films on a metal substrate was done with cyclic voltammetry (CV) technique in an aqueous electrolyte of 1 M Na_2SO_4. The CV measurement was performed in potential windows between +1 V to -1 V, as shown in Fig. 3(a-d). The voltage scan rates were varied from 5 mV/s to 50 mV/s. From the cyclic voltammetry, the specific capacitance (C_p) of TiO_2 films can be evaluated from [9]:

IOP Conf. Series: Materials Science and Engineering 1286 (2023) 012004 doi:10.1088/1757-899X/1286/1/012004

$$C_p = \frac{A}{2mk(V_2 - V_1)} \tag{1}$$

Where A is area of CV curve, m is sample mass, k is scan rate of CV measurement, $V_2 - V_1$ is the potential windows of CV measurement. The details of the calculation from the CV results in Fig. 3 are shown in Fig. 4. The results show that the prepared TiO_2 films at 10 layers exhibited optimum specific capacitance at 9.6095 F/g. Although the devices were prepared with different layers of P25-TiO_2, however, all layers are the same material. Thus, the effect of the number of layers on electrochemical properties should be the increased thickness of the film. Therefore, increasing the number of layers resulted in increasing thicknesses. Hence, the electrons could easily transport under the electrochemical measurement.

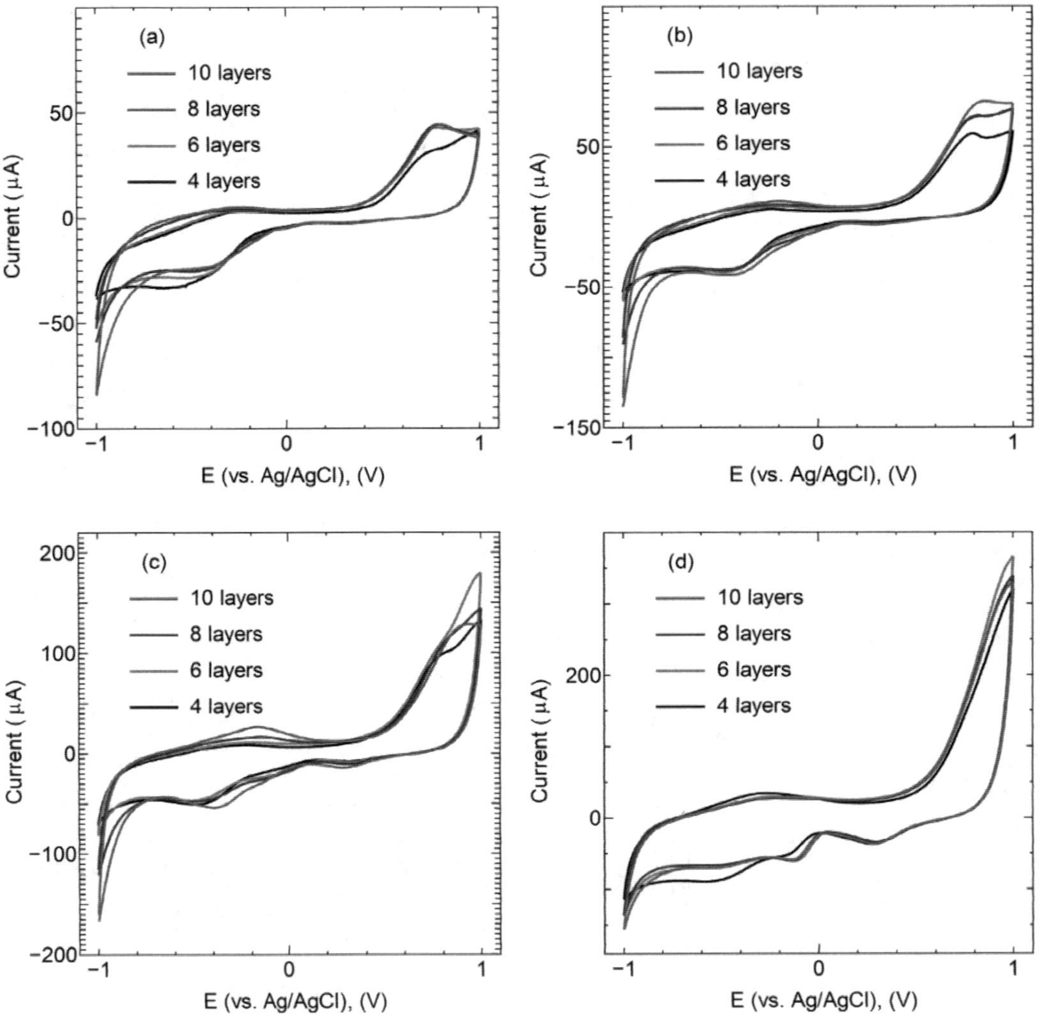

Fig. 3. Cyclic voltammogram of the prepared P25-TiO_2 films with different of voltage scan rates of (a) 5, (b) 10, (c) 20, and (d) 50 mV/s, respectively.

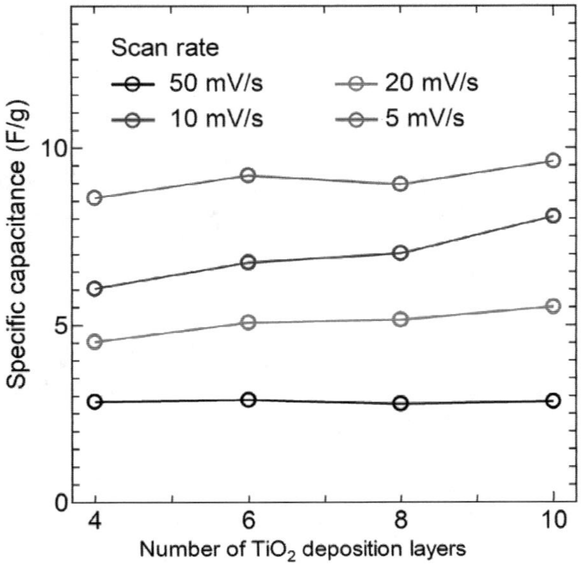

Fig. 4. Comparison specific capacitance of the P25-TiO_2 films at different deposition layers.

4. Conclusion

The titanium dioxide nanoparticle films can be prepared from commercial Degussa P25-TiO_2 nanoparticles by air-compressed spray. The prepared TiO_2 were characterized by XRD, Raman, FTIR, and SEM, respectively. The electrode of the electrochemical supercapacitor device was fabricated with air-compressed spray P25-TiO_2 on the metal substrate. The optimum specific capacitance of fabricated device was found at 9.6095 F/g at scan rate of 5 mV/seconds for prepared TiO_2 of 10 layers.

5. Acknowledgement

This work was financially supported by King Mongkut's Institute of Technology Ladkrabang (KMITL grant number: RE-KRIS/FF65/39) research fund. We acknowledge the technical assistance and facilities from NMIS of the College of Materials Innovation and Technology, KMITL

6. Declaration of Interest

The authors declare no conflict of interest.

7. References

[1] S.M. Gupta, M. Tripathi, *Chin. Sci. Bull.* **56** (2011) 1639.
[2] S. Liu, Q. Zou, Y. Ma, et al., *Inorg. Chim. Acta.*, **536** (2022) 120918.
[3] N. Ahmadi, R. Yeganeh, M. Valizadeh, et al., *Opt. Mater.,* **131** (2022) 112638.
[4] C. Prakash, and A. Dixit, *Curr. Appl. Phys.,* **41** (2022) 183.
[5] A. Ramadoss, S. J. Kim, *J. Alloys Compd.,* **561** (2013) 262.
[6] P. Ukakimaparn, D. Chantarawong, P. Songkeaw, et al., *J. Electron. Mater.* **48** (2019) 6792.
[7] J. Ma, W. Li, Benjamin J. Morgan, et al., *Chem. Mater.,* **30(9)** 2018 3078.
[8] G. Rajakumar, A. AbdulRahuman, S. Mohana Roopan, et al., *Spectrochimica Acta Part A: Molecular and Biomolecular Spectroscopy* **91** (2012) 23.
[9] N. H. N. Azman, Md Shuhazlly Mamat @ Mat Nazir, L. H. Ngee and Y. Sulaiman, *Int. J. Energy Res.*, **42(6)** (2018) 2104.

IOP Conf. Series: Materials Science and Engineering 1286 (2023) 012005 doi:10.1088/1757-899X/1286/1/012005

The Addition of TiO_2 Ratios in SiO_2–TiO_2 Nanocomposite by Sonochemical Process for the Enhancement in Self-cleaning Property

M Songpanit[1], S Limwichean[2], M Horprathum[2], W Mekprasart[1,*] and W Pecharapa[1]

[1] College of Materials Innovation and Technology, King Mongkut's Institute of Technology Ladkrabang, Bangkok, 10520, Thailand

[2]National Electronics and Computer Technology Center (NECTEC), National Science and Technology Development Agency (NSTDA), Pathum Thani, 12120, Thailand

wanichaya.me@kmitl.ac.th

Abstract. This research aims to study self-cleaning property for removing atmosphere pollutants on the substrate surface. Thus, the modification of nanocomposite material for self-cleaning property is an important challenge to solving the organic pollutants on various surfaces. Nanocomposite material in this work is focused on SiO_2-TiO_2 based metal oxide materials that synthesized by sonochemical process with the precursors of tetraethyl orthosilicate (TEOS) and titanium isopropoxide (TTIP). The crucial parameter of different amount of titanium ratios in SiO_2 matrix is related to the improvement on structural and self-cleaning properties. The ratios of Si:Ti precursor were varied at 1:0.1, 1:0.5, 1:1 and 0:1 (pure TiO_2). The structural phase of SiO_2-TiO_2 nanocomposite powder was detected by X-ray diffraction technique (XRD). Chemical bonding structure in SiO_2-TiO_2 nanocomposite was analysed by Fourier transform infrared spectrophotometer (FTIR). Surface morphology of SiO_2-TiO_2 nanocomposite was monitored by field emission scanning electron microscope (FE-SEM). SiO_2-TiO_2 nanocomposites synthesized by sonochemical process were in form of composite powder with the spherical SiO_2 particles and decorated TiO_2 on its surface. After that, the products of SiO_2-TiO_2 nanocomposite with different TiO_2 ratios were mixed with TES 40 WN solution and deposited on glass slides substrate by spray coating to study its hydrophilic property by contact angle method. The contact angle of SiO_2-TiO_2 nanocomposite film was reduced with the increase of Ti ratio in SiO_2 matrix resulting to the development of hydrophilic surface on substrate. The influence of TiO_2 in SiO_2 matrix on self-cleaning and property would be improved by hydroxyl groups on the composite surface corresponding to Ti–O–Si linkages. For self-cleaning property, Si:Ti ratio at 1:1 in SiO_2-TiO_2 nanocomposite showed the optimized condition with decreasing contact angle at 28.97 degree in 60 min after UVA irradiation.

1. Introduction

Nowadays, air pollution is one of the concerned environmental problems in a big city. Air pollution is the contaminations of dust and substances in the air that harmful to living things. To solving some problems of dust pollutants on the outside surfaces such as windowpane or glazing decoration, the addition of self-cleaning property on material surfaces is greatly interested in the presence. The mechanism of self-cleaning surfaces is divided into two different categories as hydrophobic and hydrophilic. On hydrophobic surfaces, water droplets can roll down from the surface to quickly remove contaminants due to the water repellent and low adhesive properties on the surfaces. However, the stability of the organic hydrophobic coatings under UV light irradiation needs to be improved. For hydrophilic surfaces, water droplets can spread over the entire surface in form of a water film. Thus, the contaminants on the surface may be washed away during water spreading and applied in various environmental utilizations [1]. Moreover, the advantages of self-cleaning surface are in rapid dry, anti-fogging and the elimination of light scattering due to water droplets by hydrophilic mechanism. Therefore, these evident properties with hydrophilic surface can be applied as optical devices; lenses, mirrors and biometrics [2]. For self-cleaning materials, metal oxide nanomaterials are attractively interested and improved their mechanism by light activation. In general, titanium dioxide (TiO_2) is widely used as hydrophilic materials owing to high photocatalytic activity, chemical stability and low-cost material. However, the mechanism of TiO_2 material must be excited under UV light either energy equal to or greater than 3.2 eV in TiO_2 anatase phase for producing electrons and holes pairs relating to the advanced oxidation mechanism based on hydroxyl radicals. So, UV irradiation is the main role to TiO_2 photocatalyst activation [3]. Meanwhile, silicon dioxide (SiO_2) well known as silica, has a well-defined ordered structure, large surface area, odourless, low-cost and hydrophobicity. Furthermore, SiO_2 particles are shown good cooperation with polymer matrix in the modified composite film to enhancing strength, elastic modulus and wear resistance [4]. Moreover, some research reported that the increase of hydroxyl content on SiO_2-TiO_2 composite film by sol-gel process can be improved by the surface acidity with the corporation of SiO_2 and TiO_2 resulting in the enhancement of photo-hydrophilic property [5]. Meanwhile, self-cleaning function and hydrophilic property in wool fabrics was studied through the integration of silica in the TiO_2/SiO_2 nanocomposites resulting to high capability in decomposing the stains [3]. The incorporation of SiO_2-TiO_2 nanocomposite may be improved the self-cleaning properties without UV irradiation like a natural hydrophilic [6].

Therefore, the most attractive aspect of SiO_2-TiO_2 nanocomposite is photocatalysis and hydrophilicity mechanism relating to the improvement of self-cleaning property in the material. In this work, SiO_2-TiO_2 nanocomposite was prepared by sonochemical process with the parameter of different Ti ratios in SiO_2. There is some literature reviews reported on this process. Thus, it challenged to study its formation in the system and self-cleaning property the influence of Ti in SiO_2 material.

2. Materials and Method

Tetraethyl orthosilicate (TEOS), titanium isopropoxide (TTIP), isopropyl alcohol, distilled water, absolute ethyl alcohol and oxalic acid were used as precursors and solvents in sonochemical process. The different amount of TiO_2 in SiO_2 (Si:Ti ratio) was showed in Table 1 with varied at 0:1, 1:0.1, 1:0.5 and 1:1. Firstly, SiO_2 sol was prepared by mixing absolute ethyl alcohol, distilled water and oxalic acid followed by the sonication in ultrasonic sonicator for 15 min at 750 W power supply using frequency 20 kHz and 50% amplitude. Then, TEOS solution was slowly dropped and sonicated in ultrasonic sonicator (750W, 20kHz, A 50% for 3 min. Meanwhile, TiO_2 sol was prepared by dropwise TTIP to isopropyl alcohol under magnetic stirring for 20 min. After that SiO_2 sol and TiO_2 sol were mixed under ultrasonic sonicator for 15 min at 750 W power supply using frequency 20 kHz and 60% amplitude by slowly dropping TiO_2 solution in SiO_2 sol. SiO_2-TiO_2 nanocomposite in white precipitate was washed with deionized water several times until pH 7 and dried overnight at 100 °C. For nanocomposite film, the spray coating method was applied at distance 5 cm, coating with SiO_2-TiO_2 nanocomposite at 1 %w/v were dispersed in TES 40 WN solution (commercial ethyl silicate polymer with 41% silicon dioxide) under continuous magnetic stirring and spray on glass slides substrate as prepared and baked at 100 °C overnight to obtain transparent films on glass slide. The phase identification of SiO_2-TiO_2 materials was examined by X-ray diffraction technique (XRD; Rigaku SmartLab). Chemical bonding and surface morphology of SiO_2-TiO_2 nanocomposite were monitored by Fourier Transform Infrared Spectrophotometer (FT-IR; PerkinElmer) and field emission scanning electron microscope (FE-SEM; JEOL JSM-7610F). Self-cleaning assessment was carried out by ImageJ program for contact angle measurements.

3. Results and Discussion

3.1. Physical and chemical characterization of SiO_2-TiO_2 nanocomposite

XRD patterns of SiO_2-TiO_2 nanocomposites with different TiO_2 ratio in range of 15° to 80° are shown in Figure 1. The appeared peaks in sample D pattern (pure TiO_2) at 2θ of 25°, 36°, 37°, 38°, 47°, 54°, 62°, and 69° are related to the crystalline planes of (101), (103), (004), (112), (200), (211), (204) and (220) corresponding to anatase TiO_2 phases (CSD No. 9008216). Meanwhile, the peaks at 2θ of 27° and 41° are related to the crystalline planes of (110) and (111) were identified to TiO_2 rutile phases [7]. Therefore, the product of TiO_2 nanopowder synthesized by sonochemical process was composed of mixed anatase and rutile phase. For SiO_2-TiO_2 nanocomposites in sample A, B and C, the broad characteristic diffraction peak in range of 15° to 25° is obviously occurred relating to essentially non-crystalline structure of SiO_2 [8]. Meanwhile, the characteristic peaks of mixed anatase and rutile phase in sample A, B, and C were disappeared due to the influence of a broadening SiO_2 peak [4]. Moreover, low crystallinity on broadening SiO_2 peak in SiO_2-TiO_2 nanocomposites was obviously shown by the increase of TiO_2 ratio in SiO_2 matrix due to ultrafine particles formation. These results indicated that the formation of TiO_2 phase is retarded by the presence of SiO_2 matrix in the composite form [6].

Table 1. The formula of SiO_2-TiO_2 nanocomposites with different Si:Ti ratios.

Samples	Conditions	TEOS (mol)	TTIP (mol)
A	1:0.1	1	0.1
B	1:0.5	1	0.5
C	1:1	1	1
D	0:1	0	1

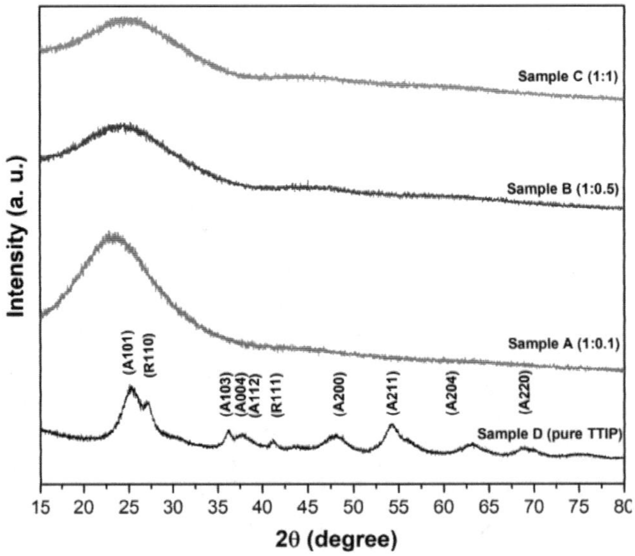

Figure 1. XRD patterns of SiO_2-TiO_2 nanocomposites with different TiO_2 ratio in SiO_2.

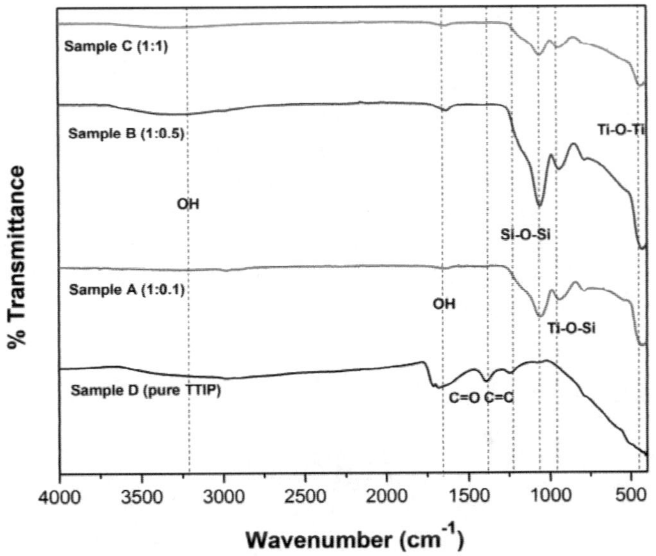

Figure 2. FT-IR spectra of SiO_2-TiO_2 nanocomposites at different TiO_2 loading.

The chemical bonding structures of SiO_2-TiO_2 nanocomposites with different Si:Ti ratios were determined by FT-IR spectrophotometer as shown in Figure 2. The identical patterns of FT-IR spectra with sample A, B and C (SiO_2-TiO_2 nanocomposites) were obtained with the different Ti precursor ratios. The intense peak at 400-800 cm^{-1} is assigned to the Ti–O–Ti bonding as identified the presence of TiO_2 phase in each conditions. Meanwhile, the apparent peaks in range of 910-960 cm^{-1} in SiO_2-TiO_2 nanocomposite spectra is related to Ti–O–Si bonding, which assigned to asymmetric stretching vibrations [9]. This result is confirmed that the formation SiO_2-TiO_2 composite is occurred by the covalent bonding of Si and Ti atoms and linking with oxygen atoms. The FTIR peaks located at 785, 812, 960, 1,075, 1,101, 1,169 and 1,392 cm^{-1} are referred to Si–O–Si bonding as specified to the presence of SiO_2 product [10]. However, the remained peaks at 1104 and 1169 cm^{-1} related to the antisymmetric stretching of Si–O–C and the rocking of the C–H bond in $-CH_3$ were corresponded to TEOS hydrolysis [11]. In the case of sample D (pure TiO_2), Ti–O stretching vibration modes was detected at 400-1000 cm^{-1} with belong to TiO_2 anatase phase. Moreover, the evident absorbance bands at 1633 and 3336 cm^{-1} are attributed to OH binding and stretching modes of adsorbed water on its surface. Some organic compounds with anatase TiO_2 particles during synthesized process as C=O, C=C bonding are observed at 1590 cm^{-1} that originated form the reaction between oxalic acid and TTIP precursors. In addition, FT-IR spectra at 3310 cm^{-1} is indicated to OH stretching of Ti–OH [12].

The microstructural features of SiO_2-TiO_2 nanocomposite powder was monitored by scanning electron microscopy with dispersive energy module are depicted in Figure 3. The morphology of sample A (Figure 3 (a)), sample B (Figure 3 (b)), sample C (Figure 3 (c)) and sample D (Figure 3 (d)) shows the difference size and shape. For SiO_2-TiO_2 nanocomposite (sample A-C), fine and spherical particles by SiO_2 formation were obviously occurred by the addition of TiO_2 loading to produce small particles. The different particle sizes were composed of SiO_2 nanoparticles in range of 83–120 nm and decorated with TiO_2 nanoparticles on its surface. The particle size of SiO_2 nanoparticles of sample A (120 nm), sample B (104 nm) and sample C (83 nm) was reduced by the increase of TiO_2 loading. While, the presence of TiO_2 particles was decorated on SiO_2 surface resulting to good particle formation and less aggregation with the increase of TiO_2 ratio in SiO_2 material as shown in Figure 3 (b) and (c). This result can be confirmed the incorporation of SiO_2-TiO_2 nanocomposite synthesized by sonochemical process corresponding to the chemical structure of Ti–O–Si bonding in FTIR spectra. For TiO_2 particles in Figure 3(d), the agglomeration of TiO_2 powder was occurred by ultrasound irradiation effect to create Ti-OH bonds from water condensation as shown in FTIR spectra [13]. This result is relating to the increase of particle collision and aggregation. Furthermore, the elemental mapping of sample C in Figure 4 revealed to the homogenous distribution of O, Si, and Ti elements on the spherical particle surface. Therefore, uniformity and good distribution of the ultrafine powder of SiO_2-TiO_2 nanocomposite are obviously obtained by sonochemical process. Meanwhile, the atomic percentages of O, Si, and Ti in SiO_2-TiO_2 nanocomposite with different TiO_2 loading in SiO_2 were confirmed by EDS technique as shown in Table 2. The increase of Ti atoms in the composite was related to the addition of TiO_2 loading in SiO_2 corresponding to the decrease of Si atoms. However, lower Ti atomic percentage in sample C was occurred due to high concentration of hydroxyl groups to form in Ti–O–Si linkages relating to the improvement in self-cleaning property [3].

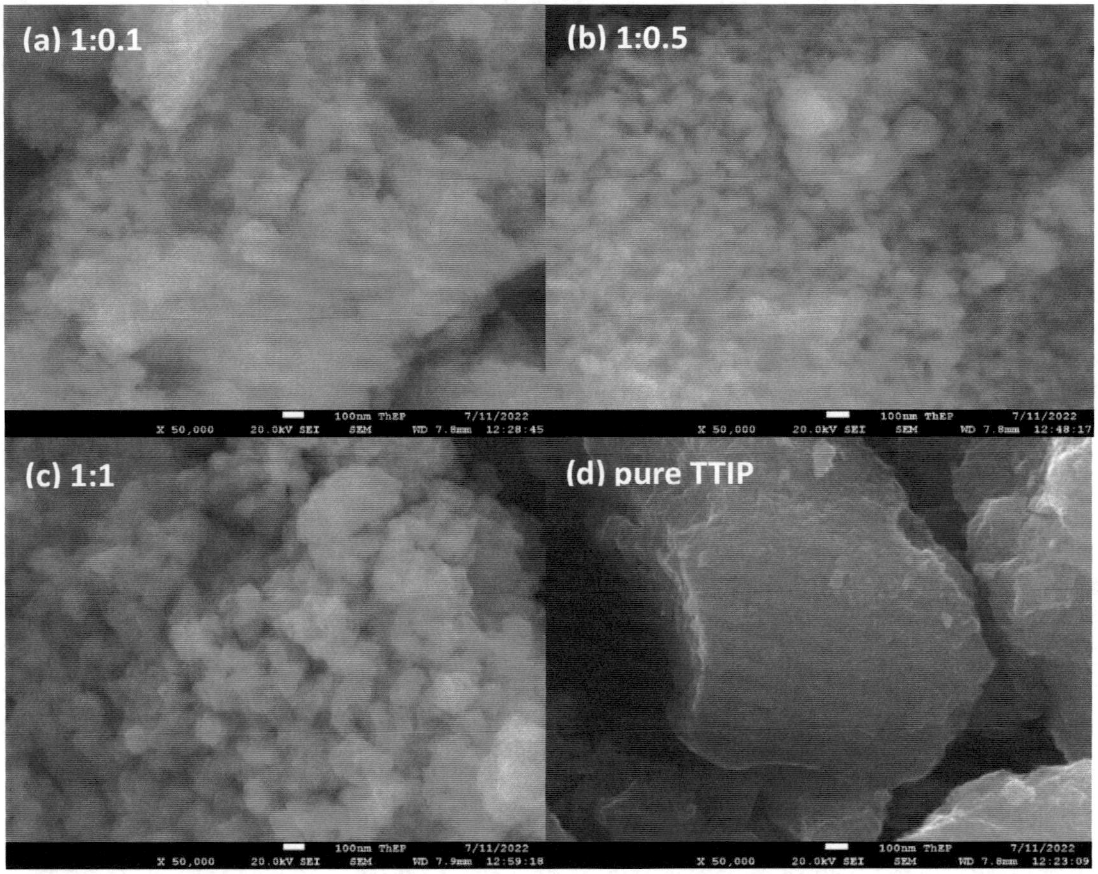

Figure 3. FE-SEM images of SiO_2-TiO_2 nanocomposite powders with (a) sample A, (b) sample B, (c) sample C and (d) sample D.

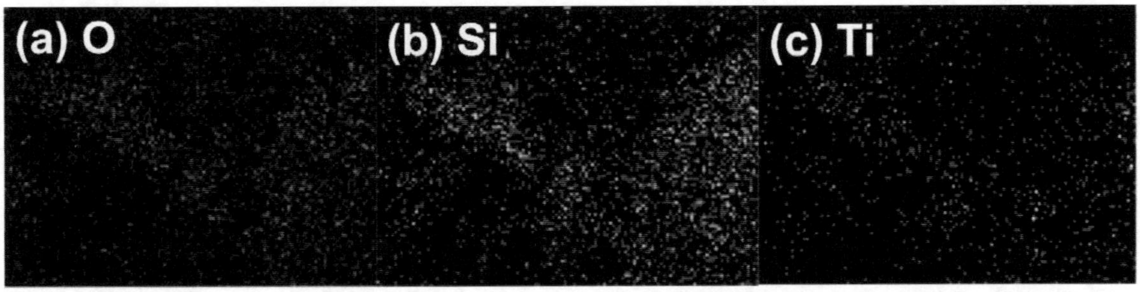

Figure 4. EDS analysis by the elemental mapping images of the SiO_2-TiO_2 nanocomposite powder (sample C) with the specific elements of (a) oxygen, (b) silicon and (c) titanium.

Table 2. Atomic percentage by EDS technique of SiO_2-TiO_2 nanocomposites with the different Si:Ti ratios.

Sample	Atomic (%)		
	Si	Ti	O
A	25.14	4.26	70.59
B	22.09	8.83	69.08
C	17.58	6.40	76.02
D	n/a	82.99	17.01

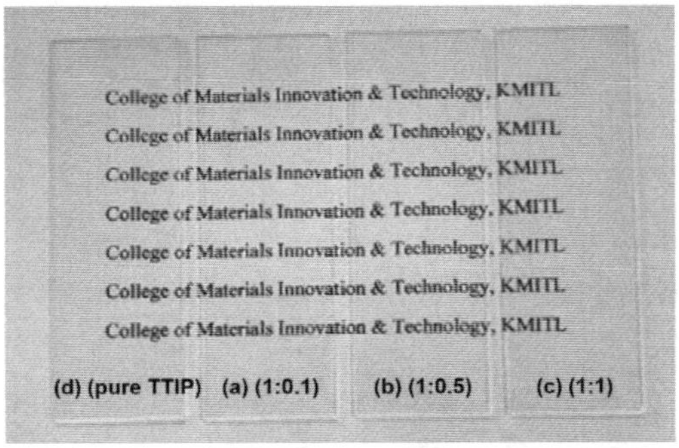

Figure 5. SiO_2-TiO_2 nanocomposite films with (a) sample A, (b) sample B, (c) sample C and (d) sample D by spray coating.

The SiO_2-TiO_2 nanocomposite films coated on glass slide substrates of sample a-d were shown in Figure 5. The transparent films were occurred after the composite coating on the glass slide substrate. The contact angle measurement by water droplets of sample A, sample B, sample C and sample D under UVA irradiation was indicated in Table 3. Before UV activation, the contact angle of water droplet on the films was obtained approximately at 60-62°. After UVA irradiation in 60 min, the hydrophilic behavior of SiO_2-TiO_2 nanocomposite films was clearly occurred by the reduction of water droplet angle due to the presence of TiO_2 phase in the composite. To increasing TiO_2 ratio in SiO_2 phase, the contact angle was decreased to 28.97° after UVA irradiation as shown in sample C

(Figure 6). Accordingly, the optimization of Si:Ti ratio in SiO_2-TiO_2 nanocomposite was at 1:1 in sample C, corresponding to the least contact angle value as self-cleaning application. The possible mechanism of the improved hydrophilic property might be originated from the increase of hydroxyl groups and oxygen vacancies on the composite surface by the presence of TiO_2 loading in SiO_2 that enhanced the polarity of hydrogen bond from water molecule in the ambient [14]. Moreover, the enhancement of hydrophilic property was occurred by the addition of TiO_2 ratios in SiO_2-TiO_2 nanocomposite by the influence of the Si–O–Ti bonding in FT-IR spectra. This result was supported hydrophilic behavior by the reduction of water droplet angle due to the presence of TiO_2 phase in the composite. Comparing with other conditions in Table 3, the contact angle measurement on the composite films was diminished under the light irradiation time. However, higher contact angle values were occurred in sample A and sample B due to less amount of TiO_2 formation in the composite [7]. Moreover, the repeatability of SiO_2-TiO_2 nanocomposite films with sample A, sample B, sample C and sample D was tested under 3 cycles by water droplet on the film surface as presented in Figure 7. The contact angle values of each condition were identical to the initial value of the first droplet cycle on the same composite film. The best performance of hydrophilic activity on SiO_2-TiO_2 nanocomposite film was also occurred at 1:1of Si:Ti ratio. These results can be interpreted that SiO_2-TiO_2 nanocomposite film was still stable and durable for the repeatability on water droplet.

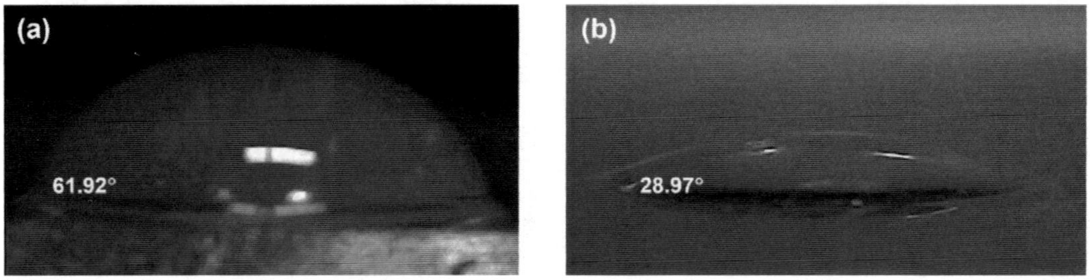

Figure 6. Contact angles of sample C (a) before UVA irradiation and (b) after UVA irradiation for 60 min.

Table 3. Contact angle measurement of SiO_2-TiO_2 nanocomposite films under UVA irradiation.

Sample	**Contact angle (°)**			
	0 mins	**20 mins**	**40 mins**	**60 mins**
A	61.89 ± 0.43	60.73 ± 0.43	59.24 ± 0.53	53.58 ± 0.45
B	63.44 ± 0.46	60.29 ± 0.81	55.97 ± 0.56	39.67 ± 0.57
C	61.92 ± 0.46	52.25 ± 0.81	45.31 ± 0.56	28.97 ± 0.57
D	62.94 ± 0.53	58.62 ± 0.51	52.91 ± 0.57	31.17 ± 0.38

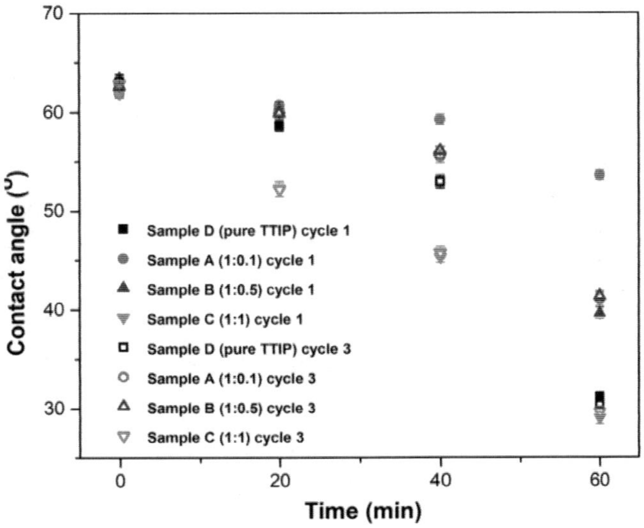

Figure 7. Contact angle measurement of the repeatability of SiO_2-TiO_2 nanocomposite films with the different of Si:Ti ratios for 3 cycles.

4. Conclusion

SiO_2-TiO_2 nanocomposites were successfully synthesized by sonochemical process for self-cleaning application with the precursors of tetraethyl orthosilicate and titanium isopropoxide. FE-SEM and EDX result confirmed that the presence of Si and Ti element in SiO_2-TiO_2 nanocomposite with homogenous powder. Meanwhile, The XRD patterns confirmed the SiO_2-TiO_2 nanocomposites by sonochemical process were non-crystalline structure as shown the broading peak in sample A, sample B and sample C. While, the mixed anatase and rutile crystalline phases in sample D (pure TiO_2) was obtained. The chemical structure of Ti–O–Si bonding in FTIR spectra was occurred in SiO_2-TiO_2 nanocomposites that related to hydroxyl groups on its composite surface. For self-cleaning property, sample C showed the optimized condition with decreasing contact angle at 29° in 60 min after UVA irradiation and stabilized film by the usage more than 3 cycles. Moreover, the morphological, chemical and self-cleaning property of SiO_2-TiO_2 nanocomposite by sonochemical process must be further studied as the variation of sonochemical parameters, high molar ratio of both precursors and nanocomposite film fabrication.

5. Acknowledgement

This work has been supported from the KMITL Research and Innovation Services (KRIS) in 2021 with the grant No. KREF016416. Meanwhile, the facility and analytical instrument have been greatly obtained by College of Materials Innovation and Technology, King Mongkut's Institute of Technology Ladkrabang (KMITL).

6. References

[1] Ren Y, Li W, Cao Z, Jiao Y, Xu J, Liu P, Li S and Li X 2020 *Appl Surf Sci.* **509** 145377
[2] Kumar A, Kashyap K, Liao K W, Hou M T and Yeh J A 2014 *IEEE.* 113
[3] Pakdel E, Daoud W A and Wang X 2013 *Appl Surf Sci.* **275** 397
[4] Teamsinsungvon A, Ruksakulpiwat C, Amonpattaratkit P and Ruksakulpiwat Y 2022 *Polym.* **14** 2729
[5] Guan K 2005 Surf Coat *Technol.* **191** 155
[6] Liu Y Y, Qian L Q, Guo C, Jia X, Wang J W and Tang W H 2009 *J Alloys Compd.* **479** 532
[7] Yazid S A, Rosli Z M and Juoi J M 2019 *JMRT.* **8** 1434

[8] Yaseen M, Shah Z, Veses R C, Dias L P, Lima E C, Reis G S D, Vaghetti J C P, Alencar W S D and Mehmood K 2017 *J Anal Bioanal Tech.* **8** 348
[9] Mohammadi F, Mirabedini S M, Khodabakhshi K and Pazokifard S 2022 *Build Environ.* **221** 109327
[10] Rodic P, Iskra J and Milosev I 2014 *J Sol-Gel Sci Technol.* **70** 90
[11] Kapridaki C, and Kalaitzaki P M 2013 *Prog Org Coat.* **76** 400
[12] Rashad M M, Shalan A E, Cantu M L and Mottaleb M S A A 2013 *J Ind Eng Chem.* **19** 2052
[13] Rosales A, Frade L O, Ramirez I E M, Godínez L A and Esquivel K 2021 *Ultrason Sonochem.* **73** 105483
[14] Du J, Wu Q, Zhong S, Gu X, Liu J, Guo H, Zhang W, Peng H and Zou J 2015 *J Rare Earths.* **33** 148

River snail shell as a highly effective renewable heterogeneous base catalyst for the production of biodiesel

Vorrada Loryuenyong, Supapisit Kooljaruswech, Supapiwat Maneenil, Surajit Duangjit, and Achanai Buasri*

Department of Materials Science and Engineering, Faculty of Engineering and Industrial Technology, Silpakorn University, Nakhon Pathom, 73000, Thailand

*Corresponding author: achanai130@gmail.com

Abstract. The biodiesel manufacturing process favors heterogeneous catalysts over homogeneous catalysts. The main drawbacks of using homogeneous catalysts are their non-renewable nature, separation, and washing, which can be avoided by using heterogeneous catalysts. This research looks into the shell of a river snail (*Viviparidae*) that has been improved with modified activated carbon (MAC) as a heterogeneous solid base catalyst for palm oil transesterification. The waste shell was repeatedly washed to remove any organic impurities attached to it and then dried in an oven. It was calcined in an air atmosphere for 2 h at a high temperature of 900 °C. The calcined sample (calcium oxide: CaO) was powdered and mixed with MAC. X-ray diffraction (XRD), scanning electron microscopy (SEM), energy-dispersive X-ray spectroscopy (EDX), and the Brunauer-Emmet-Teller (BET) method were used to characterize the CaO/MAC (mass fraction 3:1) catalyst. In order to optimize the reaction conditions for biodiesel production, operating parameters such as methanol to oil mole ratio, catalyst amount, reaction time, and microwave electrical power were investigated. As a result, the best reaction parameters were discovered to be 12:1 methanol to oil mole ratio, 2.5 wt.% CaO/MAC based on oil weight, 4 min of reaction time, and 600 W microwave electrical power. After being reused five times, the biodiesel yield could still reach 90%, indicating that the novel catalyst had good stability and recyclability. The biodiesel fuel properties obtained in this study were compared to the international biodiesel standards ASTM D6751 and EN14214. River snail shell can be thought of as a nature-based benign and resourceful material for biodiesel production, opening up a new path for fuel sustainability.

1. Introduction

The fast depletion of fossil fuels, combined with ongoing anthropogenic greenhouse gas emissions, has driven a research for alternative renewable and sustainable fuels that have efficiency similar to today's conventional fuels [1]. Biodiesel is the most well-known biofuel in the world. It is a popular substitute for petrodiesel in internal combustion engines (ICEs) due to its advantages, including non-toxicity, availability, and a similar cetane number. It can also be used directly or in blends with conventional diesel fuel without the need for modification, and it even enhances the lubrication properties of diesel fuel [2].

Because of their ease of use and low cost, homogeneous catalysts such as sodium hydroxide (NaOH) and potassium hydroxide (KOH) are now used to produce biodiesel. The reaction kinetics and mass transfer of triglycerides (TGs) in the transesterification reaction have been studied in order to optimize

IOP Conf. Series: Materials Science and Engineering 1286 (2023) 012006 doi:10.1088/1757-899X/1286/1/012006

the process conditions and reduce total biodiesel cost [3]. The use of homogeneous catalyst incurs significant operational costs, such as large wastewater processing, product purification, and catalyst neutralization, whereas heterogeneous catalyst can be easily processed, retrieved, and reused [4]. Calcium oxide (CaO) is the most widely used heterogeneous base catalyst. It is inexpensive, long-lasting, resistant to moisture, and has a low solubility in a liquid mixture. It also yields a lot of fatty acid methyl ester (FAME) [5].

Chicken bones, duck egg shells, or other types of waste shells are good sources of the raw materials required for producing CaO. The thermal decomposition of calcium carbonate ($CaCO_3$) from waste materials heated to a high temperature is anticipated to produce the conversion of calcium to CaO [6]. The catalyst made naturally from waste shells both opens the way and makes use of sustainable catalyst wastes at the same time [7]. Catalyst degradation or leaching of active metal species in the liquid phase reaction is a particular problem for the majority of solid-base catalysts. Because of this, current research has focused heavily on creating innovative CaO-based catalysts with high stability and activity, and there has been great advancement in the design of novel and effective materials [8]. By combining their expertise, the authors created modified activated carbon (MAC) supported CaO, a material in which the active ingredient and support were both generated from the shells of river snails (*Viviparidae*). The raw material utilized in catalyst synthesis provides a promising approach of creating "green catalyst" because it comes from renewable biomass and is easily biodegradable [9].

2. Experimental procedure

2.1. Materials and reagents

The river snail (*Viviparidae*) is a large operculated freshwater snail. It is classified as *Architaenioglossa*, and adults are commonly found in waterways [10]. River snail shells were collected as waste from a garden and farm in Thailand's Suphan Buri Province. The MAC used in this study was created by chemically activating palm kernel shell (PKS). The remaining substances were all analytical-grade reagents (Merck, > 99 percent purity), and they were all used in their original form.

2.2. Catalyst preparation and characterization

The PKS was washed in water to get rid of any particles or dirt, and then it was dried for 24 h at 90 °C. Using a grinder, the dried material was crushed into a 2.0 mm-sized powder and sieved. It was immersed in a potassium hydroxide (KOH) solution for 3 h at room temperature with an impregnation ratio of 2.5. Then, it was dried for a day at 90 °C in the oven. The obtained sample was subsequently heated to 750 °C for 1 h while being enclosed in a nitrogen environment [11]. The MAC was stored in glass containers after cooling.

The dried river snail shell was crushed and sieved through 60-140 mesh (105-250 μm) screens. To make the CaO/MAC catalyst, waste shell was mixed with MAC (mass fraction 3:1) and distilled water. The mixture was agitated at 80 °C for 40 min. The resultant slurry was then heated at a rate of 10 °C/min for 3 h at 900 °C in an environment of air [12,13]. All samples were maintained in a closed vessel to prevent interactions with the carbon dioxide (CO_2) and humidity in the air before use.

X-ray diffraction (XRD), scanning electron microscopy (SEM), energy-dispersive X-ray spectroscopy (EDX), and the Brunauer-Emmet-Teller (BET) method were used to characterize the CaO/MAC catalyst.

2.3. Biodiesel production

At atmospheric pressure, the transesterification process occurred within a glass reactor that was positioned inside a microwave oven and had a condenser and a mechanical stirrer. The CaO/MAC catalyst and methanol were mixed in the reactor. The heated palm oil was fed into the reactor and stirred at 600 rpm. The reaction was run at 200-600 W for 1-5 min. The mole ratio of methanol to oil was 6:1, 8:1, 10:1, 12:1, and 14:1, and the mass of heterogeneous catalysts (supporting catalyst) was 1.0, 1.5, 2.0, 2.5, and 3.0 wt.%. All tests were carried out three times, and the standard deviation for any point was never larger than 7%. The composition of the biodiesel was determined using gas chromatography-mass spectrometry (GC-MS).

3. Results and discussion

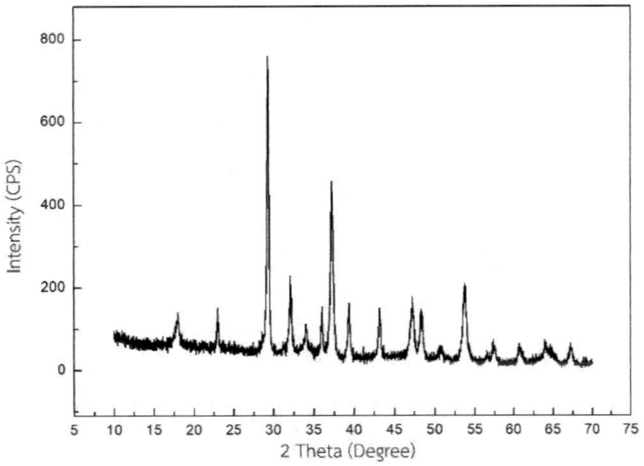

Figure 1. XRD pattern of the CaO/MAC catalyst.

Figure 1 depicts an XRD pattern of a solid catalyst. According to the pattern, the diffraction peaks that occurred at 2θ = 17°, 22°, 29°, 30°, 34°, 35°, 38°, 40°, 42°, 47°, 48°, 51°, 53°, 56°, 60°, 64°, and 67° correlates to JCPDS document number 01-085-1799, which is associated with carbon (C). The peaks associated with 2θ = 32°, 37°, 53°, 64°, and 67°, on the other hand, belong to CaO with JCPDS document number 01-077-2376. This demonstrates that the catalyst developed in the authors' investigation is primarily composed of C and CaO [9].

Figure 2. SEM image and EDX Analysis of the CaO/MAC catalyst.

The SEM image of the solid catalyst revealed a good dispersion of CaO on the MAC surface. According to these findings, activated carbon maintained its structure, which is essential for catalysis, and the calcium species was observed to be uniformly dispersed on the support's surface [14]. Figure 2 depicts the EDX spectra and elemental composition of the catalyst. The primary component elements found in the CaO/MAC catalyst were clearly calcium, carbon, and oxygen, with weight percent values of 16.32, 50.38, and 33.30, respectively.

Table 1. The physical properties of CaO, MAC, and CaO/MAC samples.

Physical property	Materials		
	CaO	MAC	CaO/MAC
BET surface area (m^2/g)	20.8	888.6	283.2
Pore volume (cm^3/g)	0.066	0.735	0.121
Pore size (nm)	4.37	0.83	1.39

Table 1 provides a summary of the physical characteristics of the CaO, MAC, and CaO/MAC samples. The solid catalyst had a pore volume of 0.121 cm^3/g and a BET surface area of 283.2 m^2/g. The BET surface area catalyst's significant reduction suggests that CaO molecules have filled the pores of the activated carbon, which supports the aforementioned hypothesis. Furthermore, it is evident that the catalyst caused a significant increase in the active sites [15]. The EDX Analysis and SEM image support this hypothesis.

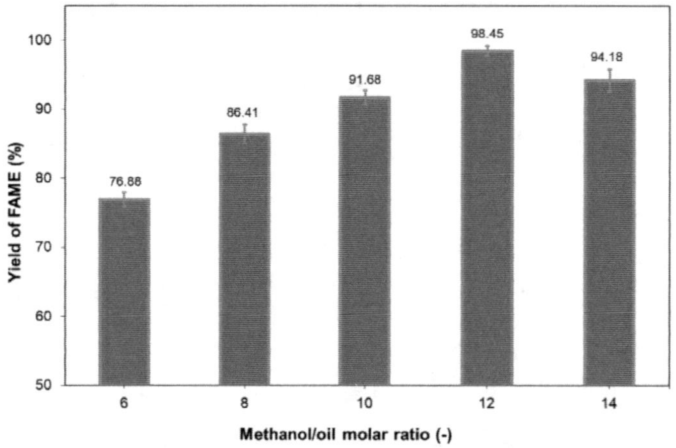

(a)

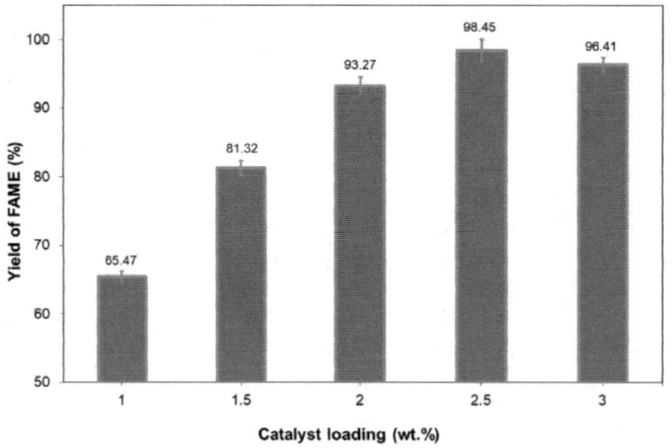

(b)

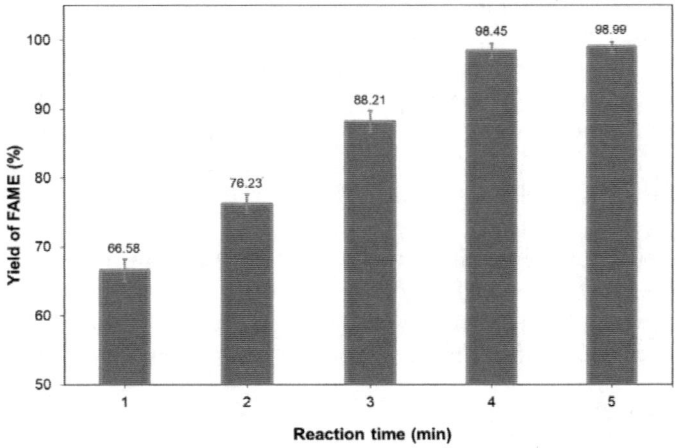

(c)

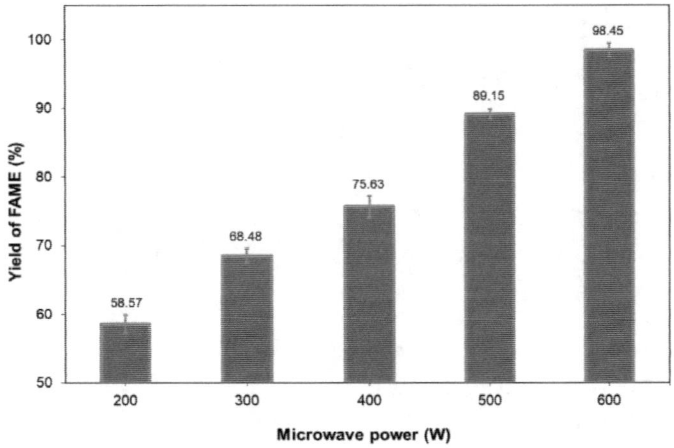

(d)

Figure 3. The effect of different reaction variables (a) methanol/oil molar ratio, (b) catalyst loading, (c) reaction time, and (d) microwave power on FAME yield (the base case is a methanol/oil molar ratio of 12:1, a catalyst loading of 2.5 wt.%, a reaction time of 4 min, and a microwave power of 600 W).

Figure 3 illustrates how a single variable, such as the methanol/oil molar ratio, catalyst loading, reaction time, microwave power, and catalyst reusability, can affect the yield of fatty acid methyl ester (FAME). In the subsequent experiments, microwave irradiation was used to catalyze the transesterification of palm oil and methanol using calcined river snail shell and MAC as the heterogeneous catalyst.

The slurry was poorly suspended in the reacting solution due to the lower methanol/oil molar ratios, which may have caused mass transfer issues and decreased catalytic activity. Higher methanol/oil molar ratios, on the other hand, resulted in a steady increase in activity (Figure 3(a)). Further boosting the methanol/oil molar ratio (12:1), though, had no positive effect on the reaction. It is recognized that in excess methanol, the glycerol would largely dissolve and prevent the methanol from reacting with the reactants and catalyst [16].

When the catalyst loading was increased from 1.0 to 2.5 wt.%, the yield of FAME climbed from 65.47 to 98.45 %. According to this finding, adding more catalyst results in increased catalyst-reactant contact, which in turn affects the rate of the reaction. However, because all of the active sites were fully occupied when more catalyst loading (more than 2.5 wt.%) was used, the yield of FAME slightly decreased [17]. In order to achieve a high yield of FAME in 4 min, it was discovered that 2.5 wt.% of river snail shell-derived CaO/MAC catalyst was the optimal amount of catalyst (Figure 3(b)).

For experiments using a heterogeneous catalyst, the reaction time, which ranged from 1 to 5 min, had a significant impact on the FAME yield. Figure 3(c) from the trials illustrates how the yield of FAME increased as the reaction time was raised from 1 to 4 min. A longer reaction time (5 min) might increase the likelihood that methanol and palm oil will come into contact with the basic sites of the catalysts, increasing the FAME yield. After 4 min, the yield slightly decreased, which could be attributed to methyl esters being hydrolyzed [18].

At different microwave powers ranging from 200 W to 600 W, the effect of the power supplied on the FAME content during the transesterification of palm oil with methanol by using CaO/MAC catalyst was examined. It was evident from Figure 3 (d) that increased microwave power resulted in a higher biodiesel yield [19]. As a result, the experiment would use a microwave power of 600 W.

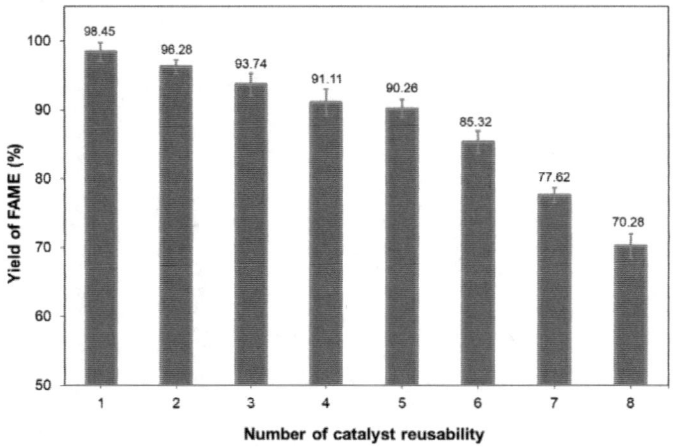

Figure 4. The effect of reusability of catalyst on FAME yield (the base case is a methanol/oil molar ratio of 12:1, a catalyst loading of 2.5 wt.%, a reaction time of 4 min, and a microwave power of 600 W).

Table 2. The fuel properties of FAME.

Fuel property	Value	ASTM D6751	EN14214	Unit
Kinematic viscosity	4.2	1.0-6.0	3.5-5.0	mm^2/s
Density	0.888	NS	0.860-0.900	g/cm^3
Flash point	160	130 Min	120 Min	°C
Cloud point	12	NS	NS	°C
Pour point	8	NS	NS	°C
Acid value	0.42	0.8 Max	0.5 Max	mg KOH/g oil
Water content	0.02	0.050 Max	0.050 Max	%

NS: not specified
Min: Minimum
Max: Maximum

A catalyst reusability test is critical because it demonstrates the benefit of using a catalyst to reduce production costs for its use in industrial applications. The reusability of the synthesized calcined river snail shell and activated carbon catalyst under optimized conditions was investigated, and the results are shown in Figure 4. The experimental results showed that an acceptable FAME yield could be obtained up to the fifth cycle. Beyond that, a significant decrease in biodiesel yield was observed, which could be attributed to pore filling and surface poisoning by triglycerides (TGs) and glycerol, resulting in a decrease in the number of active sites of the catalyst, lowering FAME yields [3].

Table 2 summarizes the prepared biodiesel properties such as kinematic viscosity, density, flash point, cloud point, pour point, acid value, and water content catalyzed by CaO/MAC catalyst synthesized from river snail shell. The results of the experiment were compared to the ASTM 6751 and EN14214 biodiesel standards. The fuel properties of FAME were consistent with the biodiesel standards [20].

4. Conclusion

The current study demonstrates the efficient use of CaO supported on MAC as a heterogeneous catalyst for palm oil biodiesel production. The prepared catalyst demonstrated promising reusability, high specific surface area, and higher FAME yield when methanol to oil molar ratio was moderate and catalyst amount was low. The following optimized reaction conditions were identified in order to achieve 98.45 percent yield FAME: reaction time, 4 minutes; methanol to palm oil molar ratio, 12:1; catalyst loading, 2.5 weight percent; and microwave electrical power, 600 W. The fuel properties of the biodiesel produced were determined using ASTM and EN standards. Biodiesel synthesis using waste river snail shells and palm oil will reduce waste disposal issues while also lowering biodiesel prices, making biodiesel a viable fuel alternative to petroleum-derived biodiesel.

Acknowledgements

The Department of Materials Science and Engineering, Faculty of Engineering and Industrial Technology at Silpakorn University is also acknowledged by the authors for its support and encouragement of this study.

Author contributions

This work was collaboratively created by all of the authors. Achanai Buasri conceptualized the experiments and wrote the first draft of the manuscript. Supapisit Kooljaruswech, Supapiwat Maneenil, and Surajit Duangjit contributed to the conceptual approach, data analysis, and results discussion. Vorrada Loryuenyong contributed to the manuscript's writing, thoroughly revised it, made critical revisions, and approved the final version.

Declaration of competing interest

The authors confirm that they have no known financial or interpersonal conflicts that might be perceived as having an impact on the research presented in this paper.

References

[1] Hui KO, Xin NK, Hwai CO, Hwei VL, Mohd SM, Yun HT-Y, Fahad AA, Abdulaziz AA and Nurul AM 2021 *Catalysts* **11** 194
[2] Samuel S, Luís N, João G, Jaime P, Rosa Q-F and João B 2019 *Energies* **12** 4670
[3] Ikbal BL, Kalyani R, Rajat G, Sushovan C, Bappi P and Samuel LR 2018 *RSC Adv.* **8** 20131
[4] Rezaei R, Mohadesi M and Moradi GR 2013 *Fuel* **109** 534
[5] Sisca V, Syukri Z and Jamarun N 2019 *Arch. Pharma. Pract.* **11(3)** 8
[6] Miskah S, T Aprianti T, Agustien M, Utama Y and Said M 2020 *AIP Conf. Proc.* **2248** 060004
[7] Hazri MM and Nasir NF 2020 *RPMME* **1(1)** 44
[8] Yanhong Z, Gang L, Zhenlv W, Jinghui S, Min Z, Wenxiang Z and Mingjun J 2010 *Energy Fuels* **24** 3810
[9] Konwar LJ, Boro J and Deka D 2018 *Energ. Source. Part A* **40** 601
[10] Kaewdaeng S and Nirunsin R 2019 *App. Envi. Res.* **41(1)** 31
[11] Buasri A, Chaiyut N, Loryuenyong V, Rodklum C, Chaikwan T, Kumphan N, Jadee K, Klinklom P and Wittayarounayut W 2012 *Sci. Asia* **38** 283
[12] Buasri A, Rattanapan T, Boonrin C, Wechayan C and Loryuenyong V 2015 *J. Chem.* **2015**

578625
[13] Shankar V and Jambulingam R 2017 *Sustain. Environ. Res.* **27** 273
[14] Baroutian S, Aroua MK, Raman AAA and Sulaiman NMN 2010 *Fuel Process. Technol.* **91** 1378
[15] Buasri A, Chaiyut N, Loryuenyong V, Rodklum C, Chaikwan T and Kumphan N 2012 *Appl. Sci.* **2** 641
[16] Buasri A and Loryuenyong V 2017 *Mater. Today: Proc.* **4** 6051
[17] Hamzah N, Yusof I, Samad WZ, Hamid HAA, Tajuddin NA and Ibrahim ML 2021 *Malays. J. Chem.* **23(2)** 11
[18] Hangun-Balkir Y 2016 *J. Chem.* **2015** 6715232
[19] Buasri A, Promsupa W, Santi Wannato S, Wanta S and Loryuenyong V 2018 *Mater. Sci. Forum* **926** 101
[20] Pandit PR and Fulekar MH 2017 *J. Environ. Manage.* **198** 319

IOP Conf. Series: Materials Science and Engineering 1286 (2023) 012007 doi:10.1088/1757-899X/1286/1/012007

Investigation on Optically Ultraviolet Response of Low-dimensional ZnO Structures Prepared by Chemical Processes

P Pintupimol, W Sinornate and W Pecharapa*

College of Materials Innovation and Technology,
King Mongkut's Institute of Technology Ladkrabang, Bangkok, 10520, Thailand
wisanu.pe@kmitl.ac.th

Abstract. Chemical processes were used to synthesize ZnO low-dimensional structure including ZnO thin film ZnO nanorod and ZnO nanoparticle. ZnO thin film was prepared by sol-gel spin coating method while ZnO nanorod was prepared by hydrothermal process and ZnO nanoparticle was synthesized via co-precipitation route. The structural, morphological, optical and photoresponse properties were investigated by XRD, FESEM, UV-Visible spectrophotometer, I-V and I-t measurement. Comparison study on photoresponse properties of all structures was conducted under UV light with UV LED wavelength 375 nm. The UV photodetector based on ZnO nanoparticle sample showed superior photoresponse time to another ZnO low-dimensional structures. The results informed crucial factors including active surface area per volume, interaction in low-dimensional structures and specific properties in each structure playing key role on optical ultraviolet response.

1. Introduction

In recent years, zinc oxide (ZnO) has emerged as a promising material for utilization in electronics, photonics and photodetectors regarding good electrical and optical properties. ZnO has a wide energy band gap of 3.37 eV [1], which is close to ultraviolet light. ZnO is an N-type semiconductor, which is very attractive to be used as a UV photodetector [2]. To enhance the photodetection capability, which is a key feature of ultraviolet photodetectors, attempts have been devoted to fabricate zinc oxide in various structures. There are several routes including decorating [3], doping [4], and surface passivation [5] to optimize and enhance its photoresponse performance. Low-dimensional nanostructures are structures that have confined directional growth less than 100 nm. Nano thin film and graphene are the examples for 2-dimensional structure while nanorod, nanowire and carbon nanotube are the group of 1-dimensional structures. Nanoparticles, nanodot and fluorene are considered as the example of 0-dimensional structure[6]. To make the devices be practical, simple synthesis process such as chemical based processes with low cost and ease of operation is considered to be comparably effective process. For photodetectors, crucial performance parameters including sensitivity, rise time and decay time have been investigated and improved. One effective way to enhance these properties has been carried out by fabrication with low dimensional materials [7].

In this work, the focus is paid on the comparison on UV photoresponse properties of low-dimensional ZnO nanostructures including thin film, nanorod and nanoparticle all synthesized via facile chemical processes.

2. Experimental

2.1. Synthesis of ZnO nanoparticles
ZnO nanoparticles were prepared by dissolving sodium hydroxide (NaOH,1 M, 65 mL) in deionized water then dropwise added into zinc nitrate solution ($ZnNO_3$,0.5 M, 100 mL) at room temperature under constant magnetic stirring for 6 h until white precipitate was deposited at bottom of beaker. The precipitate was collected by centrifuged at 5000 rpm for 10 min then the white precipitate was washed with deionized water and dried in oven at 75 °C for 6 h followed by annealing at 550°C for 2 h. ZnO nanoparticle powder was used as precursor material for high energy planetary ball milling method. 3 g of ZnO powder and 90 g of 10 mm zirconia balls were loaded in zirconia containers with milling speed at 500 rpm for 10 min[8-10].

2.2. preparation of substrate
Indium tin oxide (2.5×2.5 cm^2) was etched by aqua regia solution at middle (0.5 cm) of substrate to be an insulation zone between two conductive region as shown in Fig 1, then cleaning substrate by ultrasonic cleaner.

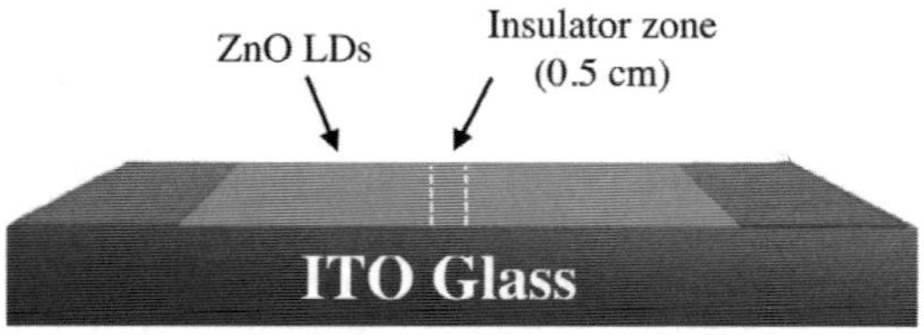

Figure 1. The schematic diagram of ZnO low-dimensional structures photodetector.

2.3. Preparation of 2-dimensional photodetector device
ZnO thin film was fabricated by sol-gel spin coating process. Zinc acetate ($Zn(CH3COO)_2 \cdot 2H_2O$, 0.2 M, 50 mL) was dissolved in ethanol, diethanolamine (($HN(CH_2CH_2OH)_2$), 99%, 1 mL) was dropped into the solution under vigorous stirring at 75°C for 2 h and then aged for 24 h. The ZnO thin film was spin-coated at 2000 rpm for 30 s followed by baking at 100°C for 10 min. The coating process was repeated 8 times to obtain designated thickness and then annealed at 500°C for 2 h.

2.4. Preparation of 1-dimensional photodetector device
ZnO nanorod was fabricated by hydrothermal process using previous ZnO thin film as a seeding layer for growth of nanorod. Zinc nitrate ($ZnNO_3$, 25 mM) was dissolved in 50 mL deionized water mixed with hexamethylenetetramine ($C_6H_{12}N_4$,25 mM) maintained at room temperature under constant magnetic string for 15 min. Flip down seed layer substrate was placed in the solution for hydrothermal process operated at 90°C for 6 h.

2.5. Preparation of 0-dimensional photodetector device
As-synthesized ZnO nanoparticle and commercial grade (UNIVAR) was coated on substrate by doctor blade technique. Polyethylene glycol 6000 MW was poured into ethanol:DI water in ratio (2:1) solution followed by loading ZnO nanoparticle powder into the solution under vigorous stirring at room temperature for 1 h. The prepared precursor was then dropped onto substrate and coated by blade then annealed at 450°C for 30 min.

2.6. Characterization of device
The crystal structures of all samples were examined by X-ray diffractometer (Rigaku SmartLab) in the range of 20 to 80 degrees. The photodetector device morphology was investigated by field emission scanning electron microscope (JOEL JSM-7001F). The optical absorption spectra of the low-dimensional structure were measured over the wavelength range of 300–800 nm using a UV/VIS-NIR

spectrophotometer (Hitachi UH4150). The current-voltage (I-V) and current–time (I–t) characteristics were measured using Oscilloscope (Rigol DS1052E) under dark and UV LED illumination (365 nm) condition. All the measurements were conducted at room temperature.

3. Results and Discussion

3.1. Structural properties

To investigate the structural features of all low-dimensional ZnO structure, the XRD spectra were measured covering the angle 2θ in the range of 20 to 80 degree. The XRD patterns of all samples illustrated in Fig 2 show that peaks at 31.75°, 34.45°, 36.25°, 47.55°, 56.60°, 62.85°, 66.49°, 67.95°, 69.10°, 72.60°, 76.95° correspond to the diffraction plane (100), (002), (101), (102), (110), (103), (200), (112), (201), (004) and (202), respectively. The results confirm the hexagonal wurtzite ZnO of all samples without any phase impurity (ICSD 2300112). Moreover, the ZnO nanorod has a (002) plane dominance due to the preferential growth direction in c-axis orientation.

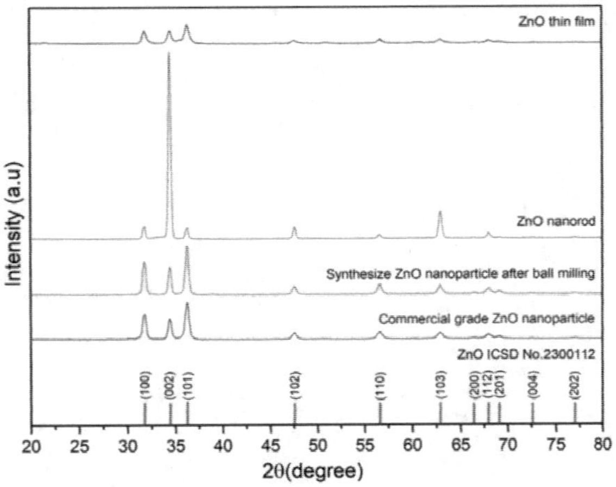

Figure 2. XRD patterns of different ZnO low-dimensional structures .

3.2. Morphology properties

To investigate the morphologies of all synthesized low-dimensional ZnO structures, the FESEM magnification ranged from 2,500X to 100,000X was carried out to observe the surface morphology and thickness of samples. Fig. 3a shows the surface morphology of ZnO nano thin film. The results exhibit dense nanoparticulate structured film with average size diameter around31 nm connected to each other. Fig. 3c exhibits the surface morphology of ZnO nanorod obtained by the hydrothermal process. The result reveals the bundle of ZnO nanorods with high aspect ratio around 56. Fig. 3e illustrates the surface morphology of a commercial grade ZnO nanoparticle with different sizes and shapes. Fig. 3g shows the surface morphology of ZnO nanoparticle synthesized by co-precipitation method revealing higher average size than that of the commercial grade ZnO nanoparticles. Fig. 3b show the cross-section images of ZnO nano thin film from spin coating process, Fig. 3d show the cross-section images of ZnO nanorod thickness from hydrothermal process, and Fig. 3f show the cross-section images of ZnO nanoparticle thickness from doctor blade technique, respectively. The corresponding results are listed in Table 1.

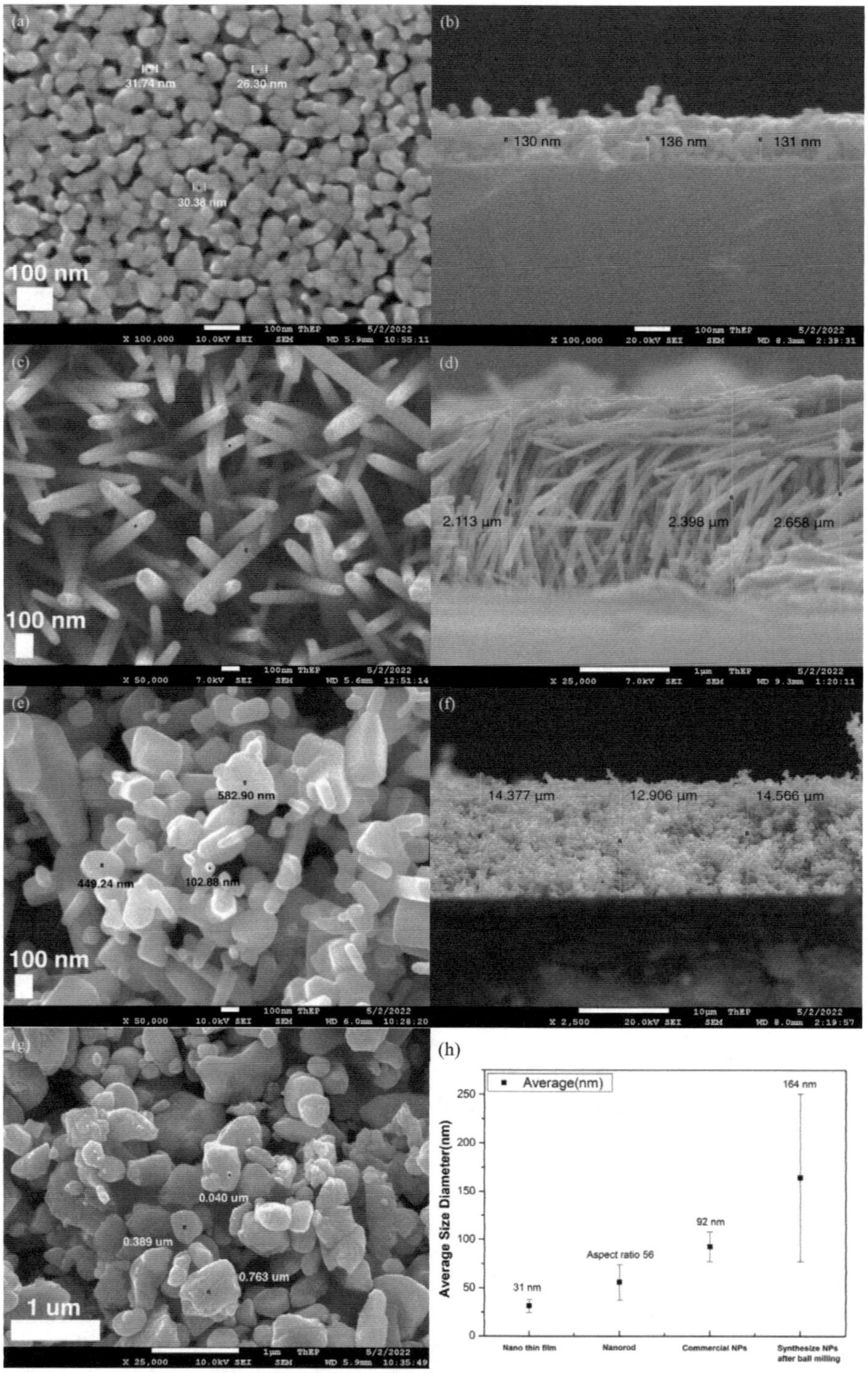

Figure 3. Morphologies from FESEM with (a) ZnO nano thin film, (c) ZnO nanorod, (e) commercial grade ZnO nanoparticle, (g) Synthesized ZnO nanoparticle. Cross section from FESEM with (b) ZnO nano thin film, (d) ZnO nanorod, (f) commercial grade ZnO nanoparticle (h) Average size of ZnO low-dimensional structures.

Table 1. Average diameter size, aspect ratio, SD and thickness of ZnO low-dimensional structure.

Sample	Diameter size	Aspect ratio	Standard deviation	Thickness
ZnO nano thin film	31 nm	-	6.76	132 nm
ZnO nanorod	-	56	18.39	2.389 μm
Commercial grade ZnO nanoparticle	92 nm	-	15.50	13.949 μm
Synthesized ZnO nanoparticle	164 nm	-	86.50	13.949 μm

3.3. Optical properties
The optical absorption spectra of low-dimensional ZnO structures were measured by UV-VIS-NIR spectrophotometer in the wavelength range of 300–800 nm. The ZnO low-dimensional structures has absorption edge around 300-380 nm corresponding to typical optical band gap of the ZnO. Synthesized ZnO nanoparticle has a lower absorption than commercial grade that is probably due to the different in the ZnO nanoparticle size. ZnO thin film has the lowest absorbance due to thinner layer of this sample. The highest absorbance is found in the ZnO nanorod that would be owing to the high long rod structure and alignment of the nanorod. These phenomena can be attributed to the multiple reflection and scattering (as proposed in Fig. 4(b)) of impinged light entering to the structure[11][12].

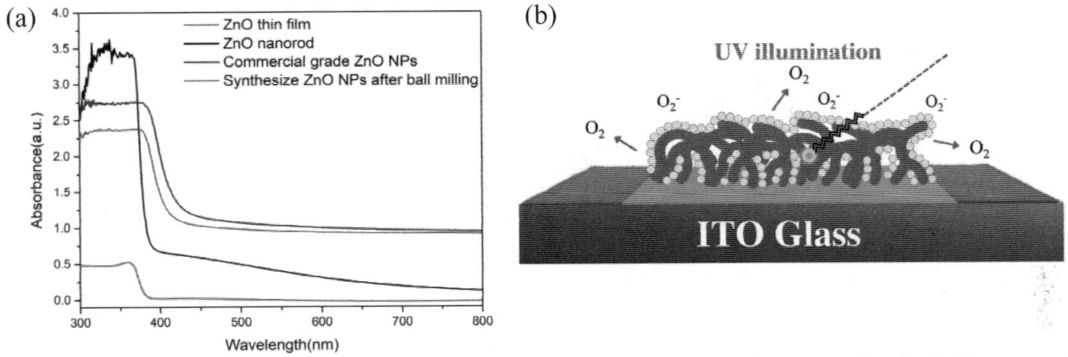

Figure 4. (a) Absorbance of ZnO low-dimensional structures, (b) Light scattering in ZnO nanorod structure.

3.4. UV responsibility
The sensitivity (R) can be calculated by following equation:

$$R = (\Omega - \Omega_0)\Omega_0^{-1}$$

where Ω_0 is resistance of the device under dark illumination, Ω is the resistance under illumination. The sensitivity is reported in the table 2. The sensitivity of commercial grade ZnO NPs exhibits superior performance comparing to the others due to fast transfer of electron in the high density of nanoparticles[13].

Fig. 5 illustrates an overview of photoresponse of all low-dimensional structured ZnO semiconductor. The shorter rise time and decay time indicate the good performance and photoresponse of the phodetector. The photoresponse versus time for all the samples under UV illumination (375 nm) on/off at the 5 V applied bias are shown. To perform this measurement, the sample was placed under UV illumination for 100 sec then under the dark condition for 100 sec.

The photoresponse mechanism in ZnO under UV illumination can be explained by the excitation of electrons on the surface of ZnO. At the beginning, oxygen molecules (O_2) in air are absorbed on ZnO surface by free electrons and the absorbed oxygen are formed to be negative charge oxygen (O_2^-) [14-16]. The negative charge oxygen on ZnO surface would effectively create a depletion layer to decrease the resistivity of ZnO [17]. When zinc oxide is excited by UV light with high photon energy, electrons from ground state are able to jump to excited state and the electron-hole pair are produced then the hole can neutralize adsorbed negative charge oxygen (O_2^-) and desorbed oxygen molecules from a surface of ZnO [18]. Electrons in excited state are allowed to move to ohmic contact electrode leading to external circuit to prevent recombination as shown in fig 6 [19][20].

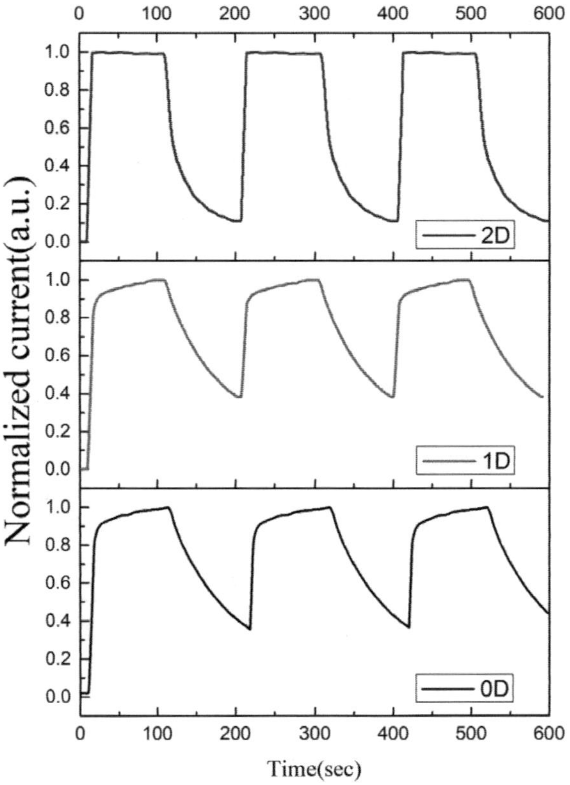

Figure 5. Overview of photoresponse in low-dimensional structured ZnO semiconductor.

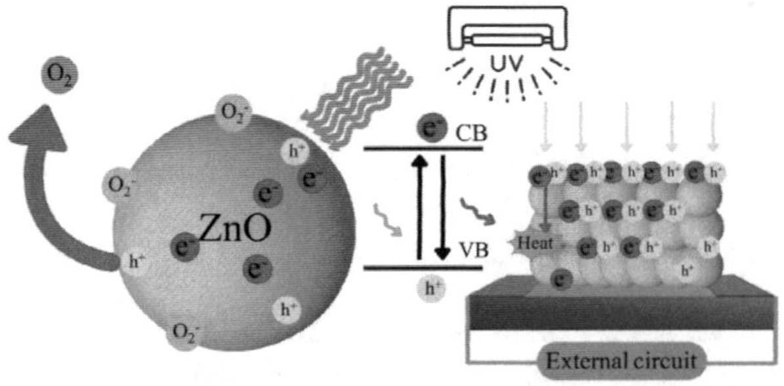

Figure 6. The generation processes of photongenerated electron-hole pairs under UV illumination, and the reactions of adsorbed oxygen on ZnO surface.

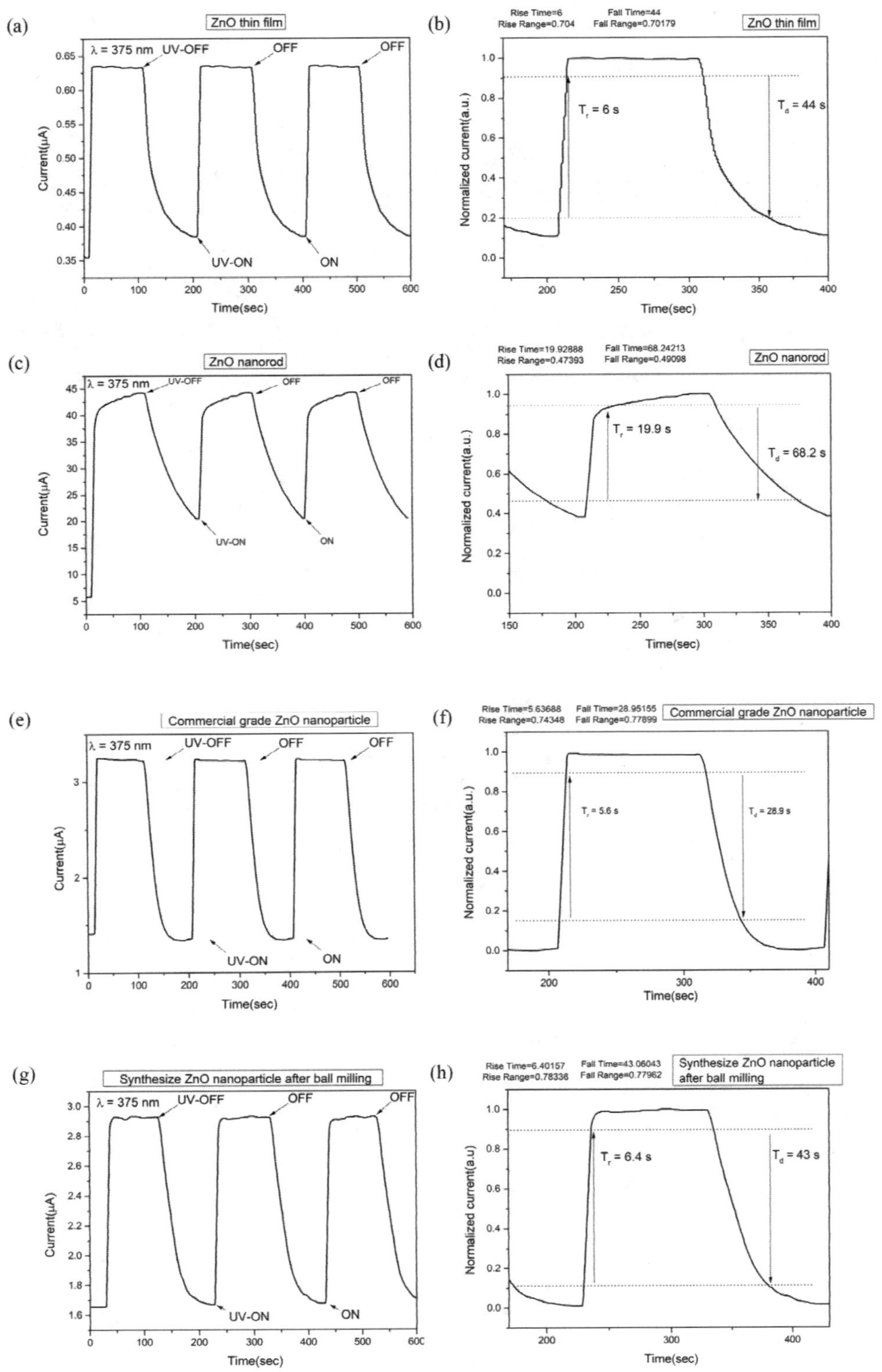

Figure 7. (a-h) Photoresponse of ZnO Low-dimenisonal structures. Time-dependent photocurrent response under 375 nm ultraviolet illumination (a), (c), (e), (g). Rise time and decay time measure from normalized response (b), (d), (f), (h).

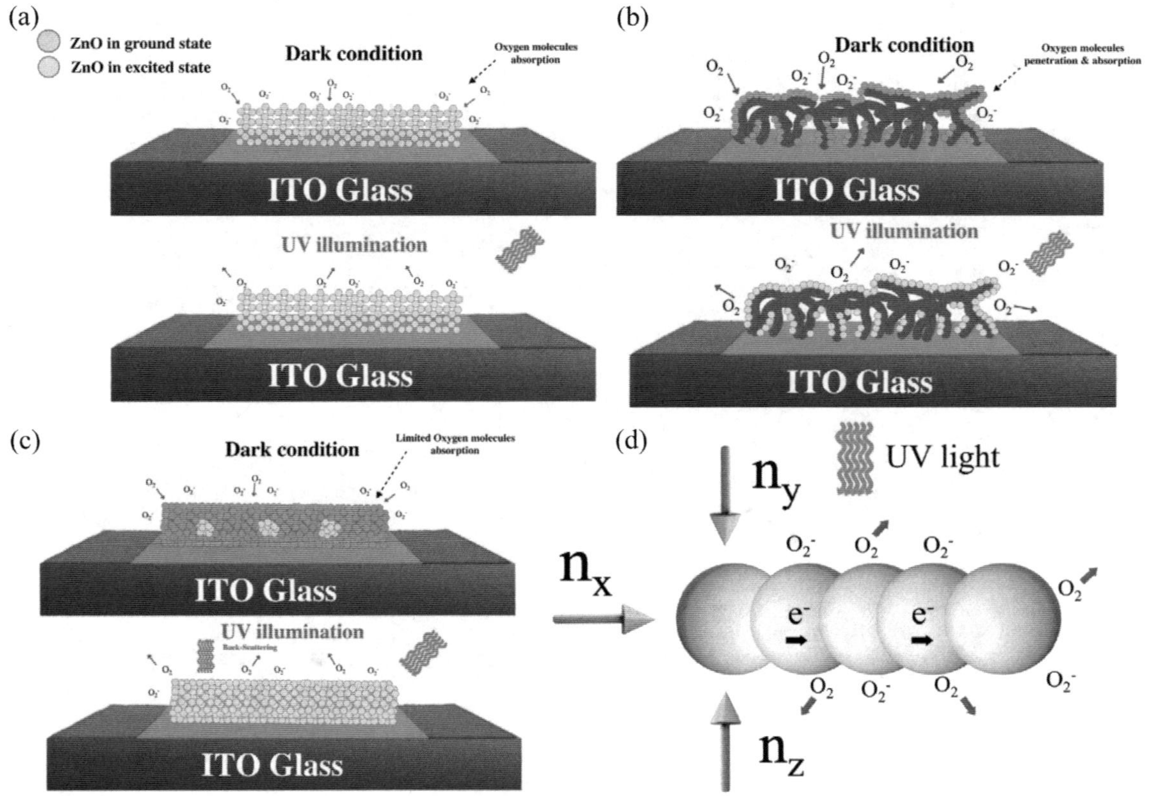

Figure 8. (a-c). Possible models for photoresponse of low-dimensional structured ZnO semiconductor (d) Fast transfer of electron caused by confined 3 dimensional of nanoparticle .

Table 2. The rise time, decay time, sensitivity, before and after illumination.

Sample	Rise time	Decay time	Before UV illumination	After UV illumination	Sensitivity
ZnO nano thin film	6.0 s	44.0 s	13.4 MΩ	2.3 MΩ	0.82
ZnO nanorod	19.9 s	68.2 s	0.292 MΩ	0.018 MΩ	0.93
Commercial grade ZnO nanoparticle	5.6 s	28.9 s	3.08 MΩ	0.17 MΩ	0.99
Synthesized ZnO nanoparticle	6.4 s	43.0 s	2.6 MΩ	0.90 MΩ	0.90

Fig. 7a-7h exhibits the photoresponse of low-dimenisonal structures ZnO semiconductor. Time-dependent photocurrent response under 375 nm ultraviolet illumination (a,c,e,g). Rise time and decay time were evaluated by mean of normalized response (b,d,f,h).

The rise time and decay time of the photodetector device were calculated and the corresponding results are listed in Table 2. Origin of longer decay time compared to the rise time of all devices can be described. ZnO nano thin film with porous structure can support and hold oxygen absorption resulting in the longer decay time than rise time. For ZnO nanorod with larger surface to volume ratio can greatly collected electron reflecting fast rise time. However, the great surfaces lead to the ease of absorbed oxygens interaction with the surface of the rods accompanying longer decay time. For ZnO nanoparticles with dense structure results in great difference between rise time and decay time due to the confinement in 3-dimensional of materials could lead to fast mobility of excited electrons whereas absorbed oxygen was limited by dense structure. Fig 8a shows a model for the photoresponse inverstigation of ZnO nano thin film. The energy is transferred from top to bottom of the device. Fig 8b illustrates a model for the photoresponse of ZnO nanorod structure investigation with large surface to volume ratio. The increased surface area resulting in greater amount of collected electron reflects to a high current around 44.25 mA. The great current would be correlated to longer rise time and decay time. Considering the device resistance, the lowest resistance was found in ZnO nanorod structure implying most conductive property. This significant feature is due to exceptional one-dimensional structure of nanorod that can improve electron transport in one direction pathwaty. Fig 8c exhibits a model for the photoresponse of ZnO nanoparticle investigation with fast rise time and decay time of 5.6 sec and 28.9 sec, respectively. This phenomenon is probably due to the dense structure, a good connection of nanoparticles that can improve the interaction between nanoparticles [21]. The confinement in 3-dimensional of materials could lead to fast mobility of excited electrons as depicted in Fig 8d. Whereas the variation in particle size could be the considerable factor to significantly decrease photoresponse performance of the device [22].

4. Conclusion

In summary, 2-dimensional ZnO thin film, 1-dimensional ZnO nanorod and 0-dimensional ZnO nanoparticle were prepared by spin coating, hydrothermal and co-precipitated process, respectively. The photoresponse performance of these nanostructures was investigated under UV illumination. It is revealed that ZnO NPs possesses superior overall performance comparing to the others with the values of sensitivity, rise time and decay time of 0.99, 5.6 s and 28.9 s respectively. To further improve the device performance, good connection of nanoparticles with good distribution in particle size are suggested.

Acknowledgement

Authors would like to express deep thank to Nanocomposite Material Research Laboratory (NMRL), College of Materials Innovation and Technology, King Mongkut's Institute of Technology Ladkrabang.

References

[1] Wang Z L 2004 *J. Phys. Condens. Matter* **16** R829-R858.
[2] Young S J and Yuan K W 2019 *J. Electrochem. Soc.* **166** B1034.
[3] Pyshkin S L, Ballato J et al 2017 London United Kingdom *Optoelectronics - Advanced Device Structures.*
[4] Qin, L., Mawignon, F. J., Hussain, M., Ange, N. K., Lu, S., Hafezi, M., & Dong, G. (2021) *Materials(Basel)* . **14,15** 4083.
[5] Hieu N M, Kim H, Kim C, Hong S K and Kim D 2016 *Journal of Nanoscience and Nanotechnology* **16,10** 10351-10355.
[6] Jasim K E 2015 *Quantum Dots Solar Cells. In (Ed.), Solar Cells - New Approaches and Reviews* p 303-311.
[7] Fatemeh A, Fahimeh Z and Mohammad Y 2021 *Optics Communications* **482** 126565
[8] Thaweesaeng N, Suphankij S, Techitdheera W and Pecharapa W 2013 *International Journal of Nanotechnology.* **11**.
[9] Samanta P 2020 *Optik* **221** 165337.
[10] Prommalikit C, Mekprasart, C, Pecharapa W 2019 *Journal of Physics: Conference Series* **1259** 012023.
[11] An J, Le T S, Lim J, Tran V T, Zhan Z Y, Gao Y, Zheng L X, Sun G Z and Kim Y J 2018 *Advanced Science* **5** 1800496.

[12] Ahmed A, Ahmed A J, Ahmed S and Naser A and Munirah A 2020 *Crystals* **10** 386.
[13] Seyedeh M, Batool S and Mohammad M 2018 *Materials & Design* **162**.
[14] Cheng J, Zhang Y and Guo R. ZnO 2008 *J. Cryst. Growth* **310** pp 57–61.
[15] Azimirad R, Khayatian A, Kashi M.A and Safa S 2014 J. *Sol-Gel Sci. Technol* **71** pp 540–548.
[16] Lim Z, Chia Z, Kevin M, Wong A and Ho G 2010 Sensors Actuators B **151** pp 121–126.
[17] Soumen D and Giri P 2011 *Nanoscale Res. Lett.* **6** p 504.
[18] Chen M W, Chen C Y, Lien D H, Ding Y and He J H 2010 *Opt. Express* **18** pp 14836–14841.
[19] Safa S, Mokhtari S, Khayatian A and Azimirad R 2018 *Opt. Commun.* **413** pp 131–135.
[20] Zhu Q. (2015) *Thickness dependent photoelectric responses of ZnO nanoparticle films*.
[21] Özgür Ü, Alivov Y, Liu C L, Teke A, Reshchikov M, Dogan S, Avrutin V , Cho S J and Morkoç H 2005 *Journal of Applied Physics* **98** 041301-041301.
[22] Nasiri N, Bo R, Wang F, Fu L and Tricoli A 2015 *Advanced materials (Deerfield Beach, Fla.)* **27**.

Preparation and Characterization of Ag Nanoparticle/SiO_2 Composite for Optical Application

Aphisit Manivong[1*], Wanichaya Mekprasart[1], Kanokthip Boonyarattanakalin[1], Krisana Chongsri[2] and Wisanu Pecharapa[1]

[1]College of Materials Innovation and Technology, King Mongkut's Institute of Technology Ladkrabang, Bangkok ,10520, Thailand

[2]Department of Applied Physics, Faculty of Science and Technology, Rajabhat Rajanagarindra University, Chachoengsao, 24000, Thailand

63607001@kmitl.ac.th

Abstract. A facile method to fabricate volumetric light-guide plate (LGP) by micro-nanostructure of Ag/SiO_2 composite was presented in this study. The Ag NPs/SiO_2 composite were first prepared from a mixture of yellowish-brown Ag NPs colloids and SiO_2 powder by electrostatic self-assembly method. Its chemical and physical properties were investigated by scanning transmission microscopy (TEM), scanning electron microscopy (SEM) and Energy dispersive X-ray spectroscopy (EDX). The small particles (Ag/SiO_2 composite) in the LGP could produce multiple scattering effect with different size of micro-nanoparticles, which could improve the light-scattering performance of the LGP. The effective light scattering of the material into LGP was obtained from the increased amount of the particles. Meanwhile, surface plasmon resonance effect of Ag NPs also improves the performance of LGP, which could be further applied to design device in optical application. In additional, optical properties and light scattering mechanism of the Ag NPs/SiO_2 composite in the LGP are discussed and presented.

1.Introdution

Nanoparticle (NPs) have been widely employed in various applications such as electronics, optical instruments, and optical fibre due to unique and exceptional optical and mechanical properties. Recently, the focus was paid on the development of optical device that called light-guide plate (LGP), which is device for applying in the backlight systems or Backlight module (BLM) [1]. Generally, LGP consists of four main parts: diffusion film (DF), light guide plate (LGP), flat reflective film (RF) and light emitting diodes (LEDs). Each component has different function in light guide plate. There are various types of LGP that have been designed from researchers to increase its efficiency [2-5]. Basically, LGP is expected to have high intensity of light from light source to manage light to the front of LGP. Various designs of LGP have been developed to meet industrial process propose. Principally, the designs of LGP have employed the same theoretical principle, the principle of light scattering, which is a principle that can be very useful in a wide variety of fields in optical devices [6-9]. There are various methods to fabricate the LGP such as replica molding method [10], lithography [11, 12], embossing process [13] and electrospray method [14-16]. However, these processes require high-end equipment to manufacture the LGP and are quite expensive due to the complex manufacturing process in the process.

IOP Conf. Series: Materials Science and Engineering 1286 (2023) 012008 doi:10.1088/1757-899X/1286/1/012008

This work aims to study and improve the optical properties of the LGP by a simple production process with low production cost. Waste material with a silica (SiO_2) content was employed to mix with epoxy resin to fabricate LGP. First, SiO_2 powder from waste glass was prepared by high energy ball milling, which is the sample process to reduces its particle size by mechanical force. For the desired application especially the light guide plate module, a specific size of embedded SiO_2 particles is strongly required to enhance light scattering and improve light emitting from the plate. The LGP was fabricated by dispersion of SiO_2 in polymer matrix. Ag NPs/SiO_2 composite was prepared and employed in LGP for further increasing performance by multi-scattering effect of the Ag NPs and SiO_2 particles.

2. Materials and Methods

2.1 Preparation of SiO_2 powder

SiO_2 powder was prepared from a wasted glass bottle (soda-lime glass) by the coarse grinding process until its particle size was less than 100 μm. After that, as-prepared SiO_2 powder was loaded in zirconia vessels and operated by a tilted planetary ball mill machine (planet M2-3F). The weight ratio of SiO_2 powder and zirconia ball (diameter 5 mm) was fixed at 1:10 in both zirconia vessels. Milling process was operated at milling speed of 500 rpm for 30 min to obtain solid powder and labelled as the first milling condition (R1). For next milling process, the prepared SiO_2 powder R1 was operated for the second milling process R2 with milling speed 300 rpm for 30 min.

2.2 Preparation of Ag NPs

60 mL of 1.25×10^{-2} M aqueous sodium citrate solution, 150 mL of 3.75×10^{-4} M aqueous silver nitrate solution and 150 mL of 5×10^{-2} M aqueous hydrogen peroxide solution was added into duran (500 mL) and then 1.2 mL of 1×10^{-3} M aqueous potassium bromide solution was added into the mixture solution in duran with magnetic stirrer 5 min. The Ag hydrosol was obtained as 75 mL of 2×10^{-2} M aqueous sodium borohydride solution was added into the mixture solution in duran with magnetic stirrer at room temperature.

2.3 Preparation of functionalized SiO_2

To make the surface of SiO_2 powder with positive charge, 0.8 g of the as-prepared SiO_2 powder from milling process were added in 500 mL ethanol and dispersed by sonication, and then 4 mL of APTES was added in with magnetic stirrer 5 h at 60 °C. After that, the functionalized SiO_2 was washed with ethanol and deionized water for removing the residual and dried at 100 °C in oven for 12 h.

2.4 Preparation of Ag NPs/SiO_2 composites

The Ag NPs/SiO_2 composites was obtained by electrostatic self-assembly method. In detail, 0.1 g of SiO_2 modified with APTES was added into 200 mL DI water, then adding 10-50 mL of the Ag colloids (0.1 mg. mL^{-1}) with magnetic stirrer for 1 h. The Ag NPs/SiO_2 composites powder was washed with deionized water and dried at 100 °C in oven for 12 h. [17]

2.5 Preparation of Light guide plate with scattering particle

Light guide plate was prepared from epoxy resin. In a typical synthesis process, 6.0 g of bisphenol A $(CH_3)_2C(C_6H_4OH)_2)$ and the concentrations of relative particles (SiO_2 and Ag/SiO_2) with different wt% were mixed with magnetic stirrer 15 h at room temperature. Then, 3 g of hardener $C_9H_{12}N_2O$ was added to the precursor solution and mixed well by stirring for 10 min. The Light guide plate with a thickness of 0.25 mm was obtained by pouring into molding target (6 cm x 6 cm). Finally, it was dried at room temperature for 24 h.

3. Results and Discussion

3.1 Characterization of SiO_2 and Ag NPs for become scattering particle into diffusion film

The waste material for this work is soda-lime glass, which is a type of glass with SiO_2 as the main constituent. Therefore, the elemental composition of the powder from soda-lime glass milled by high energy ball milling process was quantitatively determined by XRF. Table 1 shows the major SiO_2 compound of milled powder of 68.69%. This result suggests that the milled powder can be used as scattering particles in diffusion film because SiO_2 is the transparent material, which can cause the incident light to change the direction as called "Light Scattering". It can occur when light travels in transparent solids or liquids. Light scattering phenomena in light-guide plate have two major mechanisms including Rayleigh scattering and Mie scattering. Rayleigh scattering is the elastic scattering of light by particle with particle size much smaller than the wavelength of incident light and Mie scattering is the scattering when the size of the scattering particle is equal to or larger than the wavelength of incident light. In addition, the intensity of scattered transmitted light could be determined by equation 1.

$$I_s = I_0\left(\frac{8\pi^4 N a^6}{\lambda^4 r^2}\right)\left|\frac{m^2-1}{m^2+1}\right|^2 (1+cos^2\theta) \quad (1)$$

Wherein, I_0, N, a, θ, r and m are incident light intensity, the number of particles, particle size, the scattering angle, the distance of the observation points from the scattering center and the ratio of the refractive index of the scattering center to that of the medium (epoxy resin and scattering particle), respectively.

Table 1. XRF measurement of waste materials milled by high energy ball milling

Content of Elements (%)											
SiO_2	Na_2O	CaO	MgO	Al_2O_3	SO_3	Cl	K_2O	TiO_2	Fe_2O_3	Y_2O_3	ZrO_2
68.69	11.39	14.14	1.97	1.28	0.24	0.04	0.15	0.11	0.30	0.09	1.64

In our work, the scattering particles were designated by SiO_2 microparticle and Ag NPs to produce Rayleigh scattering and Mie scattering when light travels inside the light diffusion film. Table 2 shows the particle size of SiO_2 powder as prepared by high energy ball milling and Ag as prepared by simple redox chemistry.

Table 2. The particle size analysis for SiO_2 powder and Ag NPs

Sample	Particle size (nm)			Average
	1	2	3	
SiO_2	715.2	771.6	888.4	791.7
Ag	88.0	91.3	84.0	87.8

The mechanism of light scattering when light travels into epoxy (light diffusion film) that have scattering particle inside. In figure 1. shows schematic of (a) SiO_2 and (b) Ag/SiO_2 dispersed in epoxy resin. When light travels in (a), the incident light is scattered as Mie scattering by SiO_2 microparticle, the light will travel forward rather than sideways. In (b) the light at sideways direction can be increased via both Mie scattering and Rayleigh scattering by Ag/SiO_2 composite. Scanning electron microscope (SEM) images of SiO_2 and Ag/SiO_2 composite powder composite are shown in Figure 2 (a) and (b). Uniform dispersion of Ag NPs on surface of SiO_2 can be clearly observed in scanning transmission electron microscopy (TEM) images are shown in Figure 2 (c) and (d). The elemental mapping in Figure 2 (e) further verifies that Ag NPs (Red dot) distribution on the surface of SiO_2 microparticle (Si = Green and O = Blue). The Si, O and Ag elements can be identified in energy dispersion spectra (EDS) in Figure 2 (f).

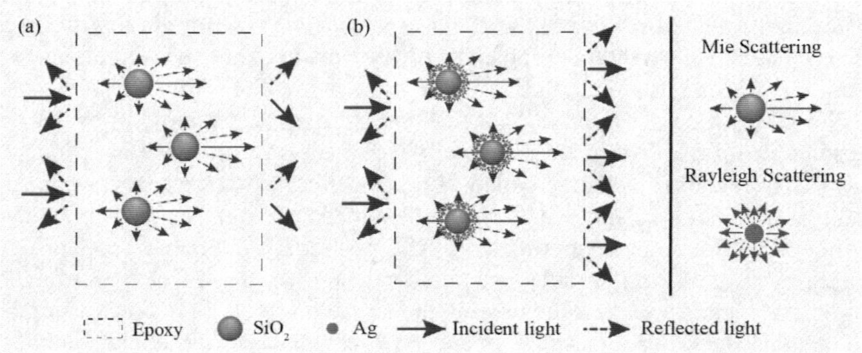

Figure 1. Schematic of light scattering in light diffuser film as (a) contain with SiO_2 and (b) contain with Ag NPs/SiO_2 composite.

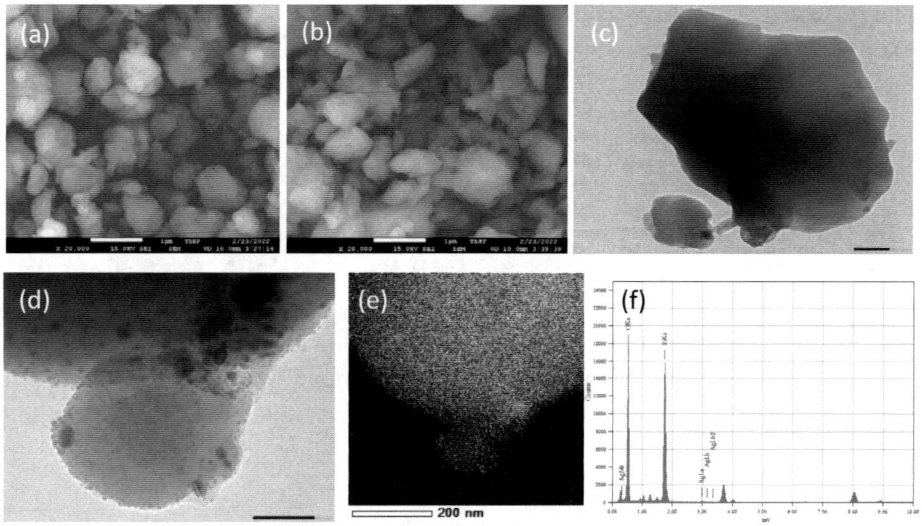

Figure 2. SEM images of (a) SiO_2 powder as prepared by high energy ball milling and (b) Ag/ SiO_2 powder as prepared by electrostatic self-assembly method. (c) TEM image of SiO_2, (d) TEM image of Ag/ SiO_2, (e) elemental mapping and (f) Energy Dispersion Spectra (EDS) of 1% Ag/ SiO_2.

3.2 Optical properties of the light diffuser based on the scattering particle as fillers

The optical transmittance spectra of light diffuser films fabricated by epoxy resin as fillers scattering particle shown in Figure 3 (a) for SiO_2 powder and (b) for Ag/ SiO_2 composite powder. The results indicate that the light transmittance spectra of epoxy resin is over 87% and then decreases with increasing scattering particle into epoxy because the missing direct light was scattered in other directions. Moreover, the transmittance of the light diffuser films slightly decreases with the increasing amount of Ag as the surface of SiO_2. This result is suggested that the increasing Ag NPs could increase the light absorption more than light scattering because the Ag NPs could be aggregated to each other leading to the increase in particle size.

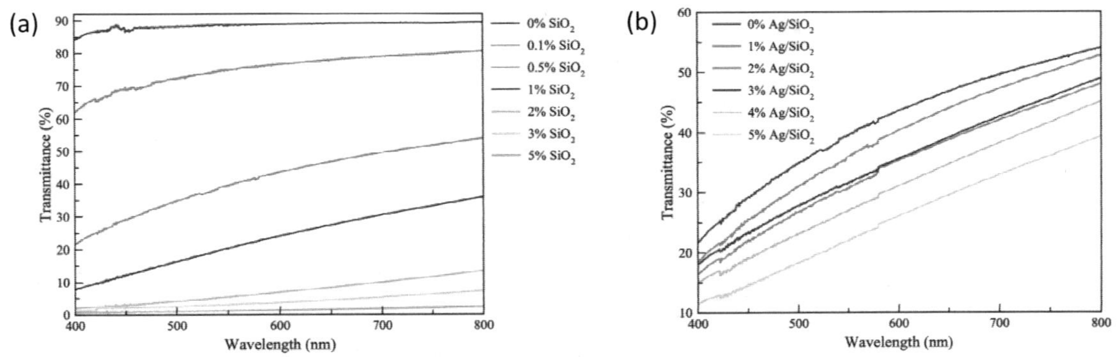

Figure 3. UV-Visible spectra of light guide plate (a) containing different amounts of SiO_2 and (b) containing 0.5% of Ag/SiO_2 as different contents of Ag.

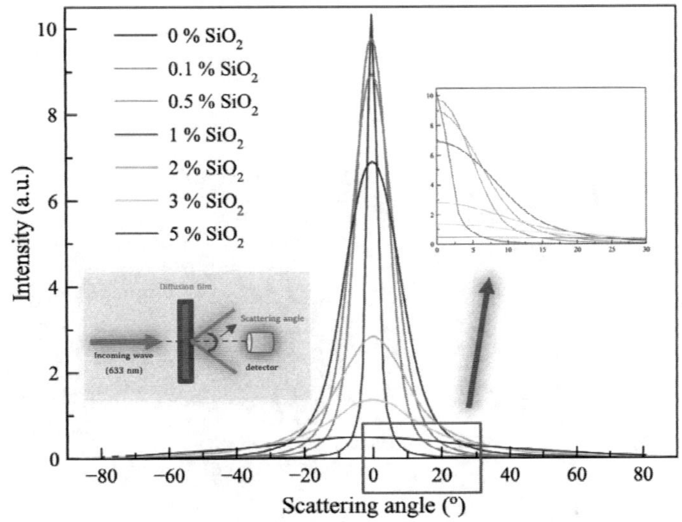

Figure 4. The scattering profiles (wavelength = 633 nm) of light diffusion film containing different amounts of SiO_2.

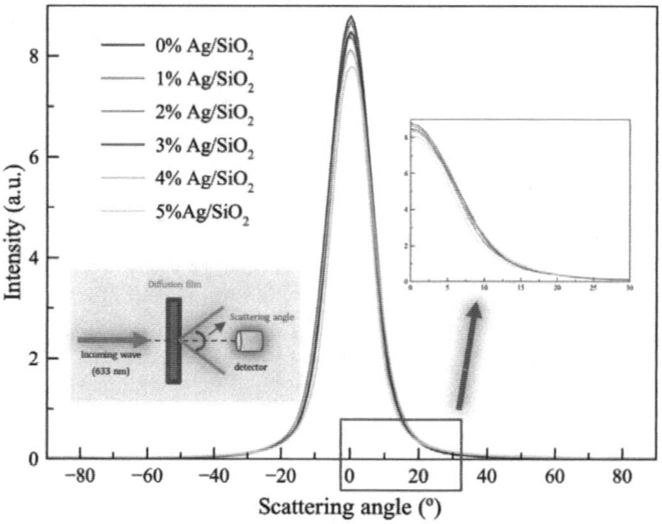

Figure 5. The scattering profiles (wavelength = 633 nm) of light diffusion film containing 0.5% of Ag/SiO_2 as different contents of Ag.

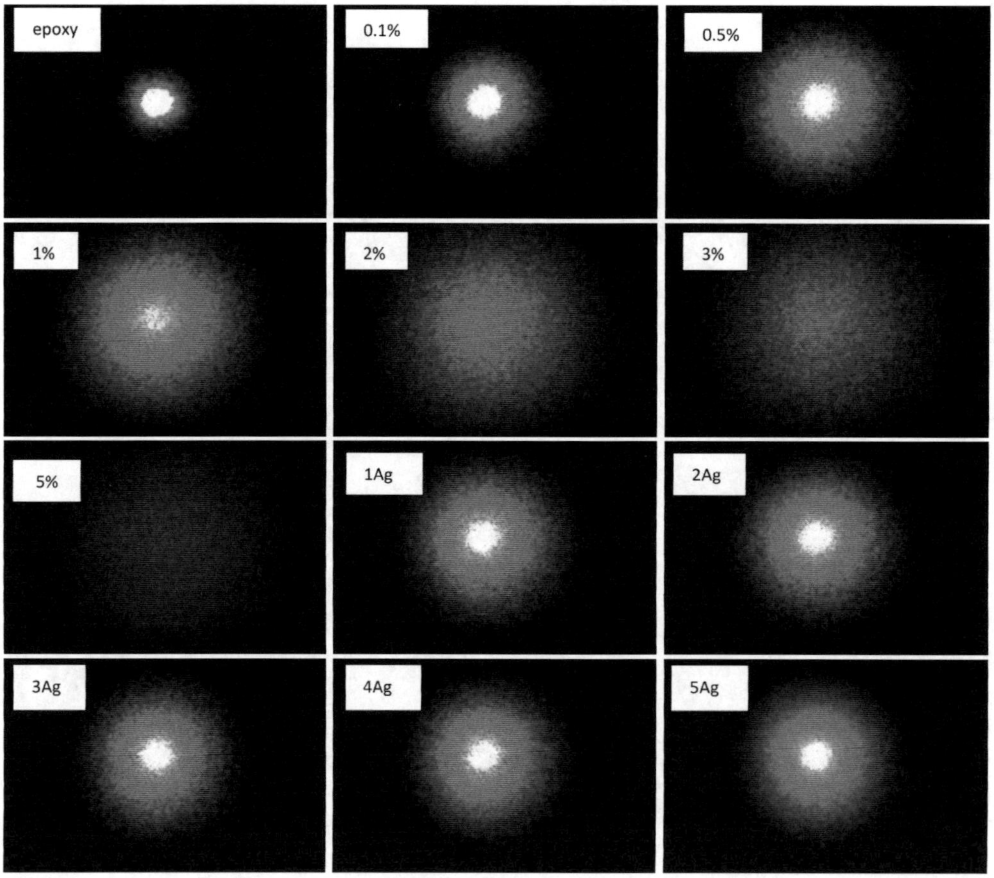

Figure 6. Photo of light transmittance form laser 633 nm thought (a) epoxy light diffuser film with 0.1, 0.5, 1, 2, 3, 4, 5 %wt. SiO_2 powder and with 1, 2, 3, 4, 5 %wt. Ag/SiO_2 composites.

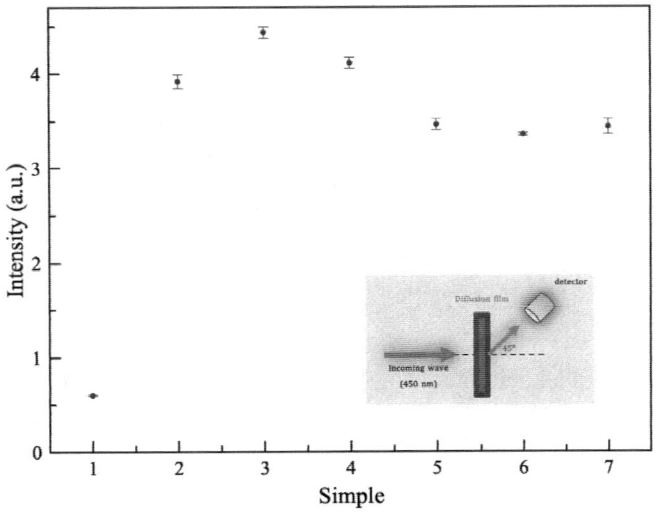

Figure 7. The scattering profiles (wavelength = 450 nm) at 45 degrees with different simple of light diffusion film as (1) epoxy, (2-7) 0%, 1%, 2%, 3%, 4% and 5% of Ag/SiO_2 composite, respectively.

The particle size and number of particles are a key role in the transmittances and homogenization of the incident light of the diffusion film. The light scattering on the surfaces of diffuser film that have scattering particles and have significantly influence on the enhancement of the diffuse transmittance. As seen in Figure 4, the scattered light of the diffuser film increases with the increasing number of scattering particles. Meanwhile, the direct transmittance of the diffuser film decreases with the increasing number of scattering particles. In addition, the diffuser film with Ag/SiO_2 composite in Figure 5 shows slightly increasing light scattering and light absorption of Ag NPs. The homogenization of the light penetrated throughout the diffuser films is shown in the scattered light image in Figure 6. When comparing the scattered light under different wavelength of light shown in figure 7, it reveals that the Ag NPs could affect to the light scattering of diffuser film as different wavelength when Ag NPs are dispersed on the surface of SiO_2.

4. Conclusion

In summary, the light diffuser films based on SiO_2 as fillers were facilely fabricated into epoxy resin at room temperature. The SiO_2 powder was prepared by high energy ball milling process to reduce its size to micro range for using as scattering material to cause Mie scattering inside diffuser film. Moreover, Ag/SiO_2 composite was constructed to increase the efficiency of the light diffuser film, which could improve the light scattering efficiency by multiple scattering effect (Mie scattering and Rayleigh scattering). Additionally, light scattering of diffuser film increases when the particle loading was increased. Meanwhile, the scattering angle of light was also changed depending on the size of particle and the wavelength of light. Thus, SiO_2 and Ag/SiO_2 composite has potential to be used as scattering particles in application as light diffuser films.

Acknowledgement

This research grant has partially been provided by the Program Management Unit: PMU B (Grant No. B05F630019). Authors would like to thank College of Nanotechnology, King Mongkut's Institute of Technology Ladkrabang (KMITL) for supporting synthesis and characterization facilities.

References

[1] Huang Y, Zhang X, Yang W, Ke H, Li C, Wang X and Xu P 2021 *Optik* **239** 166624
[2] Yao P-H, Chung C-J, Wu C-L and Chen C-H 2012 *Opt. Express* **20** 4819-29
[3] Xu P, Huang Y, Zhang X, Huang J, Li B, Ye E, Duan S and Su Z 2013 *Opt. Express* **21** 20159-70
[4] Wang Y-J, Ouyang S-H, Chao W-C, Lu J-G and Shieh H-P D 2015 *Opt. Express* **23** 1567-75
[5] Chen E, Xie H, Huang J, Miu H, Shao G, Li Y, Guo T, Xu S and Ye Y 2018 *Opt. Express* **26** 3466-82
[6] Kuo H P, Chuang M Y and Lin C C 2009 *Powder Technology* **192** 116-21
[7] Hu J, Zhou Y, He M and Yang X 2013 *Optical Materials* **36** 271-7
[8] Bhunia A 2008 *Advances in food and nutrition research* **54** 1-44
[9] Mudalige T, Qu H, Van Haute D, Ansar S M, Paredes A and Ingle T 2019 *Nanomaterials for Food Applications,* **pp** 313-53
[10] Yi Wu C, Chiang T H and Hsu C C 2008 *Opt. Express* **16** 19978-86
[11] Chen M F, Pan C T and Hwang Y M 2011 *Sensors and Actuators A: Physical* **167** 130-8
[12] No K, Kim G and Kim G M 2008 *Journal of Mechanical Science and Technology* **22** 1765-71
[13] Liu C F, Pan C T, Liu K H, Chen Y C, Chen J L and Huang J C 2011 *Journal of Materials Engineering and Performance* **20** 1544-53
[14] Wang J-H, Chen C-F, Ho J-R, Shih T-K, Chen C-C, Whang W-T and Yang J-Y 2009 *Optics & Laser Technology* **41** 804-8
[15] Ahn S and Kim G 2009 *Applied Physics A* **97** 125-31
[16] Kim G H and Park J H 2007 *Applied Physics A* **86** 347-51
[17] Hou Z, Chu J, Liu C, Wang J, Li A, Lin T and François-Xavier C P 2021 *Chemical Engineering Journal* **415** 128863

Humidity Sensor Based-on Quartz Crystal Microbalance of Sprayed Silver Nanoparticles Films

Benchapol Tunhoo[1], Arnon Wechayan[1], Savita Churasri[1], Thutiyaporn Thiwawong[1,*] and Korakot Onlaor[1]

[1] Electronics and Control System for Nanodevices Research Laboratory, College of Materials Innovation and Technology, King Mongkut's Institute of Technology Ladkrabang, Bangkok, 10520, Thailand.

*E-mail: thutiyaporn.th@gmail.com

Abstract. The silver nanoparticles (Ag NPs) film was prepared in this work by the electrostatic spray deposition technique. The Ag NPs film properties were characterized by X-ray diffraction, UV-visible spectrophotometer, Raman spectroscopy, and scanning electron microscope. Then, the Ag NPs film was prepared on a 2 MHz crystal microbalance to fabricate the humidity sensor. The measurement systems are composed of a fabricated sensor, oscillator circuit, and frequency counter. The sensor performances were obtained in terms of the frequency shift of the device at various humidity levels. The conditions of saturated salt solutions following NIST standard were used as humidity values for measurement. It was found that the fabricated device exhibited approximately a frequency shift of 25 Hz in the humidity range of 11-93 %RH. Moreover, the device demonstrated low hysteresis at a value of about 1%RH with a humidity level of 43 %RH.

1. Introduction

Humidity is an important physical environment parameter. Generally, the humidity value is represented in terms of saturated moisture vapor in the surrounding air. The influences of humidity in the environment are serious to consider for determining and controlling in various applications such as pharmaceutical, agriculture, and electronics device. Therefore, the humidity sensing device is required with high sensing characteristics, high stability, high response time, and low hysteresis error [1].

Silver nanoparticle (Ag NPs) is an interesting nanomaterial in the field of nanotechnology[2]. Due to the properties of physicochemical and good antimicrobial behaviors, the Ag NPs were used to apply in various fields of research and applications. Ag NPs had been performed in biomedicines, textiles, food packaging, cosmetics, etc. In those numbers, the application in electronics devices of Ag NPs had been attractively studied, such as sensor [3], conductive filler [4] and memory device [5], etc.

The quartz crystal microbalance (QCM) is a mass sensing device with a gas-phase reaction. This system consisted of an AT-cut plane piezoelectric disk composed of electrodes at the top and bottom of the disk. The oscillator circuit was connected to QCM device and generated the frequency signal output due to the piezoelectric properties of quartz material and mechanical oscillations of the crystal [6]. The frequency output of QCM depended on the change of mass of QCM following the Sauerbrey equation [7].

In this work, the Ag NPs films were prepared by electrostatic spray deposition technique on the QCM. The physical properties of prepared films were investigated. The humidity sensing characteristics of the prepared device were performed.

2. Experiment

This work used the commercial 30 wt % of 50 nm silver nanoparticles (Sigma Aldrich) in ethylene glycol as a precursor solution. The electrostatic spray deposition (ESD) method was applied to prepare films. The ESD experiment setup and prepared film are illustrated in Fig. 1(a). The Ag NPs films were deposited on a 2 MHz quartz crystal microbalance (QCM) substrate at various deposited times, as shown in Fig. 1(b). The deposition time varied by 5, 10, 15, and 20 seconds. In case of longer deposition time, the wet surface on the prepared films was found, which cannot prepare the device at room temperature. The physical characteristics such as crystalline structure, chemical structure, surface morphology, and optical absorbance of Ag NPs films were performed by X-ray diffraction (SmartLab, Rigaku), Raman spectroscopy (DXR SmartRaman, Thermo Scientific), scanning electron microscope (EVO 15, Zeiss), and UV-Visible spectrophotometer (T92+ Spectrophotometer, PG Instrument), respectively. Then, the prepared films on QCM were used for the humidity sensor. The humidity sensing characteristics of the device were performed with the oscillating circuit with an inverter gate (74LS04) and two 4.7 KΩ resistors, as shown in Fig. 1(c). The frequency of the oscillating circuit can be measured with a frequency counter (Digicon FC-2700) at various humidity levels. The saturated salt solutions were used as humidity source for sensing measurement with the relative humidity of 11, 23, 32, 43, 52, 75, 84, and 93%RH by using the saturated aqueous solutions of LiCl, CH_3COOK, $MgCl_2$, K_2CO_3, $Mg(NO_3)_2$, NaCl, KCl, and KNO_3, respectively. The sensor characteristics, such as sensitivity, hysteresis, and response times, were evaluated from the measurement data.

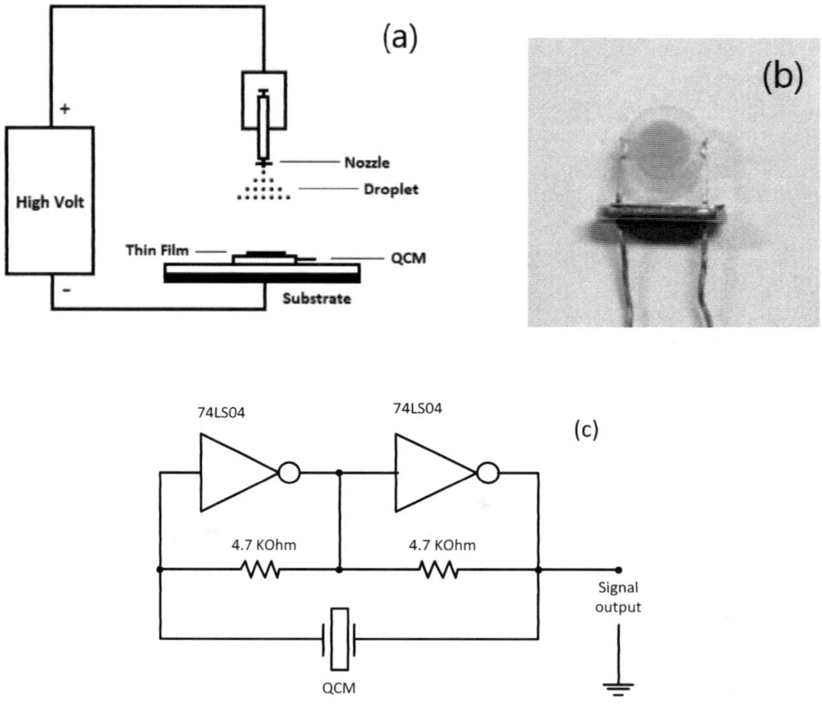

Figure 1. (a) The electrostatic spray deposition setup, (b) a quartz crystal microbalance (QCM) sensor, and (c) the QCM oscillating circuit schematic measurement.

3. Result

The XRD spectra of prepared Ag NPs are shown in Fig. 2(a). The spectra exhibited diffraction peaks at a 2-theta angle of 38 degrees that corresponded with the crystalline orientation planes of (111). With increasing deposition times, a higher diffraction peak intensity can be observed. While the full width at half maximum of diffraction peak decreased. It was indicated to increase the crystallinity in deposited films. In addition, the subtle peak was observed at a diffraction angle of 44 degrees with an orientation plane of (200). That result was consistent with other reported results.

Fig. 2(b) depicts the Raman spectra of Ag NPs films. It was found to peak at wavenumber about 244, and 465 cm^{-1}. The characteristic of Raman band at 244 cm^{-1} was due to the stretching mode of Ag-O bonding [8]. That causes the formation of oxygen and silver nanoparticles during the deposition films process. Moreover, the Raman peak at 465 cm^{-1} was defined as the stretching vibration of C-N-C [9]. It might be due to polymer encapsulation or the stability agent of silver nanoparticles.

Optical absorption spectra of Ag NPs films in the region of 350 to 800 nm are shown in Fig. 2(b). The strong absorption peak at about 440 nm corresponds with the surface plasmon characteristics of Ag NPs. Those results confirmed the formation of Ag NPs films on the substrate prepared by the ESD technique. Moreover, the absorption peak was shifted to a longer wavelength (red shift) due to the effect of the reciprocal diameter of Ag particles [3].

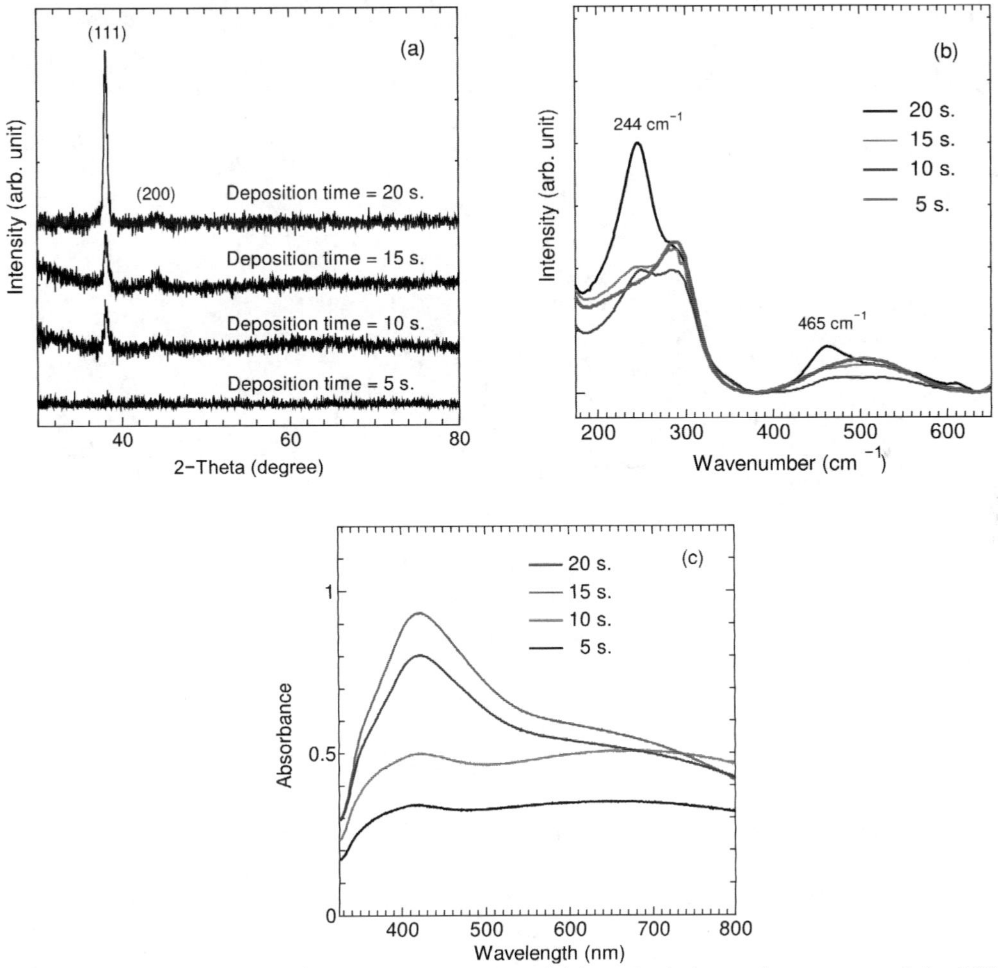

Figure 2. (a) XRD spectra (b) Raman spectra, and (c) optical absorption spectra of Ag NPS at different deposition times.

IOP Conf. Series: Materials Science and Engineering 1286 (2023) 012009 doi:10.1088/1757-899X/1286/1/012009

Surface morphologies of prepared Ag NPs films at different deposition times are shown in Fig. 3. The prepared film at a short deposition time, such as 5 seconds, exhibited a cluster of nanoparticles covering the surface. When increasing the preparation time, the surface morphologies illustrated the dense particle films over the substrate surface. Due to in this work, the heat treatments such as post thermal dry/annealing process was not used in prepare process; therefore, the morphology of the prepared was composed of the main morphology of Ag NPs. The fine nanoparticles on the morphology of all prepared films can be seen. Moreover, the wet surface over the deposition films was observed at longer deposition time.

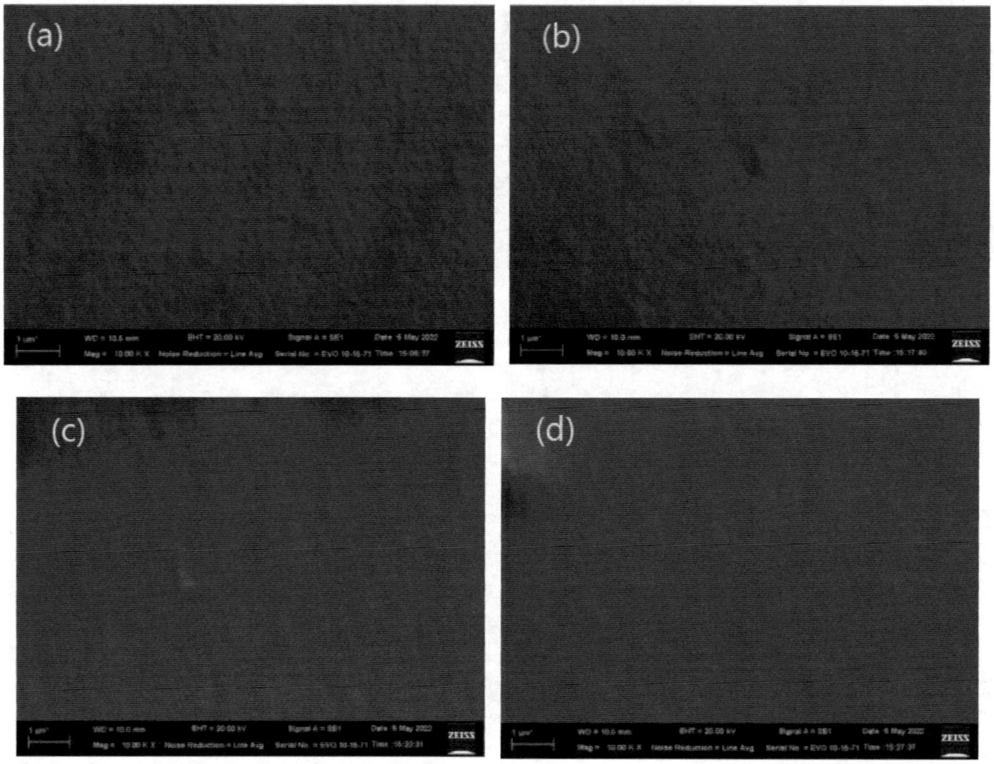

Figure 3. Surface morphologies of Ag NPs films prepared at (a) 5, (b) 10, (c) 15, and (d) 20 seconds.

To obtain the humidity sensing characteristics of Ag NPs films on QCM, the frequency of prepared QCM devices was performed with the 11%RH for 600 seconds as based-line data. Then, the device was moved to each vessel of saturated aqueous salt solutions and recorded the frequencies. The results of frequency shifts of Ag NPs QCM sensor are shown in Fig. 4(a). The humidity sensitivity (S) was defined as the relation of frequency shift (Δf) in the range of the relative humidity (Δ%RH) as:

$$S = \frac{\Delta f}{\Delta \% RH} \tag{1}$$

It was found that the QCM device with a deposition time of 20 seconds exhibited optimum humidity responses.

Hysteresis is an imbalance of adsorption and desorption processes during humidity measurement [10]. The adsorption and desorption processes are based on capturing and releasing moisture molecules with the sensing layer. The hysteresis (H) of Ag NPs QCM device can be calculated following [11]:

$$H = \frac{\Delta f_{D43} - \Delta f_{A43}}{S} (\% RH) \tag{2}$$

When S is sensitivity, Δf_{D43} and Δf_{A43} are the frequency shift values at a humidity level of 43%RH of desorption and absorption processes, respectively. It was found that the hysteresis values of Ag NPs QCM devices prepared at 10, 15, and 20 seconds are lower than 1%RH. The maximum hysteresis value was found in the Ag NPs QCM device with a preparation time of 5 seconds with a value of about 1.15%RH. Those values exhibited lower hysteresis than typical commercial values [11].

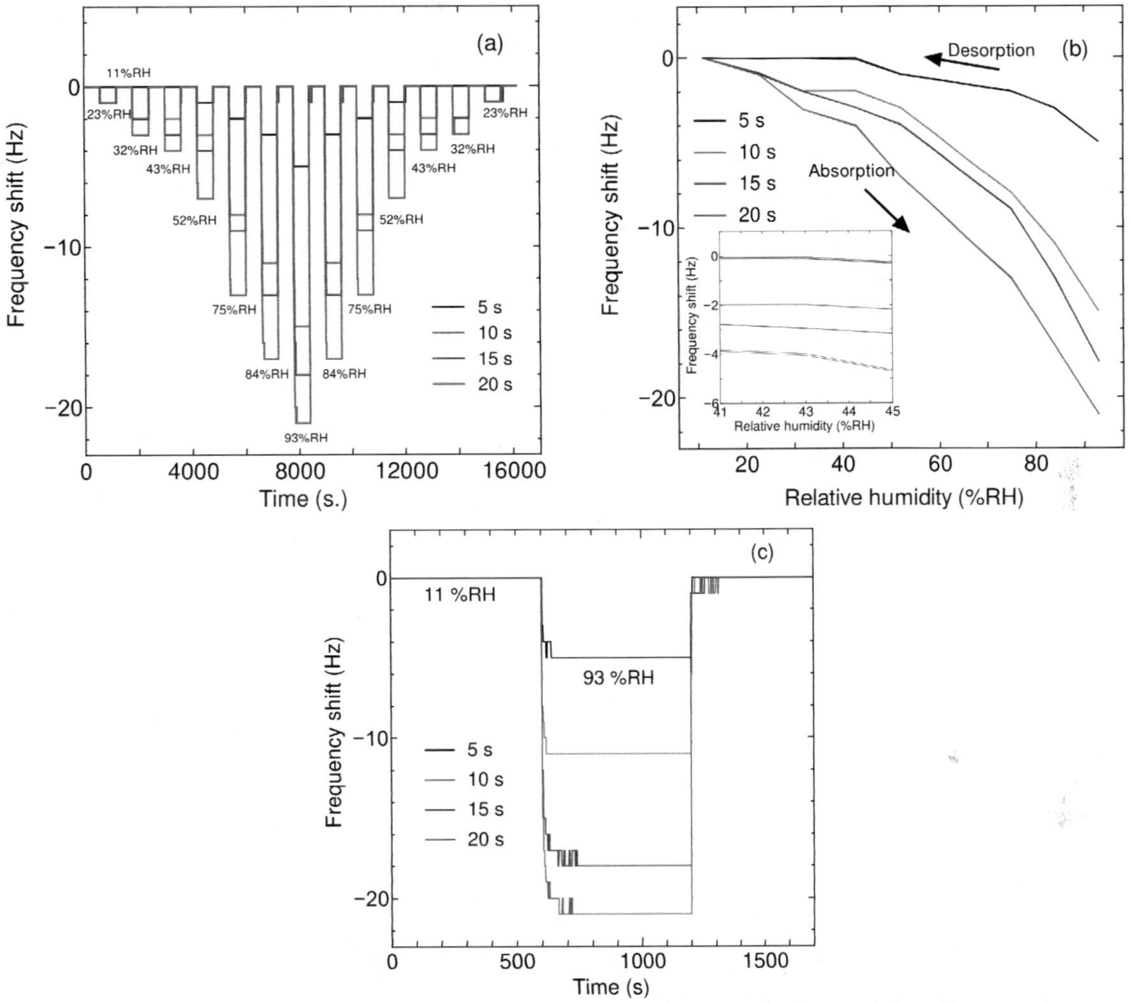

Figure 4. (a) Humidity response, (b) hysteresis characteristics, inset: the zoon-in of hysteresis curves in humidity range between 41 – 45 %RH, and (c) cyclic time response of between humidity levels of 11%RH – 93 %RH of Ag NPs QCM device at various deposition times.

The transition time from 10% to 90% time of change was used to perform the response time. The Ag NPs QCM device's response times can be measured by measuring the frequency shift in two levels of humidity value at 11%RH and 93%RH, as shown in Fig. 4(c). The humidity levels had changed rapidly from 11%RH → 93 %RH → 11%RH. The absorption and desorption times of prepared Ag NPs QCM device can be achieved from time of change of frequency shift from humidity levels of 11%RH to 93%RH, and 93%RH to 11%RH, respectively. Summarized humidity sensing characteristics, including sensitivity, adsorption, and desorption times of prepared Ag NPs QCM are shown in table 1.

Table 1. Summarized characteristics of Ag NPs QCM humidity sensor.

Deposition time (s)	Sensitivity (Hz/%RH)	Adsorption time (s)	Desorption time(s)
5	5.53×10^{-2}	17	5
10	1.72×10^{-1}	7	4
15	2.05×10^{-1}	24	4
20	2.56×10^{-1}	13	5

4. Conclusion

The 2MHz quartz crystal microbalance (QCM) with deposited silver nanoparticles films by electrostatic spray deposition was applied to prepare the humidity sensor. The effect of deposition duration times on the physical properties of Ag NPs films were performed by X-ray diffraction, Raman spectroscopy, and UV-visible spectrophotometer, which indicated the nanostructure of Ag NPs films. The fabricated Ag NPs QCM humidity sensor demonstrated the humidity sensitivity of 2.56×10^{-1} HZ/%RH for preparing Ag NPs films for 20 seconds. Moreover, the Ag NPs QCM humidity sensor exhibited low hysteresis with high response absorption and desorption times.

Acknowledgement

This work was financially supported by King Mongkut's Institute of Technology Ladkrabang (KMITL grant number: RE-KRIS/FF65/39) research fund. The authors acknowledge the facilities and technical assistance from Nanotechnology and Materials Analytical Instrument Service Unit (NMIS) of College of Materials Innovation and Technology, King Mongkut's Institute of Technology Ladkrabang.

Declaration of Interest

The authors declare no conflict of interest.

References

[1] Chen Z and Lu C 2005 Humidity sensors: a review of materials and mechanisms *Sensor letters* 3 274-95

[2] Islam M A, Jacob M V and Antunes E 2021 A critical review on silver nanoparticles: From synthesis and applications to its mitigation through low-cost adsorption by biochar *Journal of Environmental Management* **281** 111918

[3] Thiwawong T, Onlaor K and Tunhoo B 2013 A humidity sensor based on silver nanoparticles thin film prepared by electrostatic spray deposition process *Advances in Materials Science and Engineering* **2013**

[4] Syafiuddin A, Salmiati M and Salim A 2017 Beng Hong Kueh, T. Hadibarata and H. Nur *J. Chin. Chem. Soc* **64** 732-56

[5] Onlaor K, Thiwawong T and Tunhoo B 2019 Flexible and fully transparent WORM memory devices based on Ag nanoparticles blended with poly (ethylene-co-vinyl acetate) *Synthetic Metals* **258** 116200

[6] García-Martinez G, Bustabad E A, Perrot H, Gabrielli C, Bucur B, Lazerges M, Rose D, Rodriguez-Pardo L, Fariña J and Compère C 2011 Development of a mass sensitive quartz crystal microbalance (QCM)-based DNA biosensor using a 50 MHz electronic oscillator circuit *Sensors* **11** 7656-64

[7] Erol A, Okur S, Comba B, Mermer Ö and Arıkan M 2010 Humidity sensing properties of ZnO nanoparticles synthesized by sol–gel process *Sensors and Actuators B: Chemical* **145** 174-80

IOP Conf. Series: Materials Science and Engineering 1286 (2023) 012009 doi:10.1088/1757-899X/1286/1/012009

[8] Joshi N, Jain N, Pathak A, Singh J, Prasad R and Upadhyaya C P 2018 Biosynthesis of silver nanoparticles using Carissa carandas berries and its potential antibacterial activities *Journal of Sol-Gel Science and Technology* **86** 682-9

[9] Arvizo R and Bhattacharyya S 2012 R. a. Kudgus, K. Giri, R. Bhattacharya and P. Mukherjee *Chem. Soc. Rev* **41** 2943

[10] Chen Q, Feng N-b, Huang X-h, Yao Y, Jin Y-r, Pan W and Liu D 2020 Humidity-sensing properties of a BiOCl-coated quartz crystal microbalance *ACS omega* **5** 18818-25

[11] Liu H sensitivity and low-hysteresis porous MIM-type capacitive humidity sensor using functional polymer mixed with TiO2 microparticles *Sensors* 0284

Preparation of Copper Oxide Films by Electrochemical Deposition Technique for Extended Gate Field Effect Transistor pH Sensor

Thutiyaporn Thiwawong[1], Kotchakorn Lertnawanin[1], and Benchapol Tunhoo[1,*]

[1] Electronics and Control System for Nanodevices Research Laboratory, College of Materials Innovation and Technology, King Mongkut's Institute of Technology Ladkrabang, Bangkok, 10520, Thailand.

*E-mail: benchapol.tu@kmitl.ac.th

Abstract. In this work, copper oxide (CuO) film was prepared by electrochemical deposition technique on commercial indium-tin oxide/glass substrate with two electrode configuration. The precursor solution was the 0.1 molar aqueous of copper sulfate. The films were prepared at different times of 30 to 120 second. Then, the deposited films were annealed at 450 °C to achieve CuO film. The influence of electro-deposition voltage and deposition times on structural property, morphological features had been studied. The CuO films were used to fabricate the pH sensor based on extended-gate field effect transistor (EGFET). The sensitivity and linearity of prepared device was performed in the standard buffer solutions with pH range of 2-12. It was found that the device demonstrated the sensitivity and linearity of 28.5 mV/pH and 90.1 %. It can be seen that the prepared CuO films can be used as pH sensing application with EGFET device.

1. Introduction

Metal oxide nanostructures have interested to use in widely applications due to the advantage properties such as low cost, simple process of synthesis etc. Copper oxide (CuO) is a one of metal oxide family that widely researched in many field of usages [1]. CuO is a p-type semiconductor monoxide with a narrow bandgap. CuO exhibited more advantage properties such as nontoxic, high stability, and low cost of material.

The pH value is important parameters of liquid that effected on many applications such as environmental, biological, chemical process etc [2]. Generally, the pH measurement can be achieved by various sensing methods such as glass electrode, ion-sensitive field effect transistor, light-addressable potentiometric sensor, extended-gate field effect transistor. One of those, extended-gate field effect transistor or EGFET has received attractive to study and research. Due to the facile structure of EGFET that can be fabricated of sensing layer for connect to the commercial MOSFET device. This configuration was made separation between sensing layer and MOSFET device, which established less influence of environment parameters such as operation temperature, light illumination, and disposable sensing device. Normally, the main sensing of EGFET was based on the surface potential which was related to the charge accumulate of protonation or de-protonation process on the surface of sensing layer. The modulated surface potential between electrolyte and sensing layer resulted on the electrical characteristic of commercial MOSFET [3].

In this work, the CuO films were prepared by electrodeposition process with various deposition times and voltages. The physical properties of prepared films were performed. The prepared films were applied to sensing layer in extended gate field effect transistor (EGFET) for pH sensor. The sensing characteristics of CuO films were investigated

2. Experiment

In this work, the electro-deposition process was applied to prepare copper oxide films. The precursor solution was an aqueous 1 M CuSO4. The deposited film was prepared on the commercial indium tin oxide (ITO) on glass substrate. The anode and cathode electrodes for electro-deposition process were composed of ITO substrate and gold wire. Prior, the substrate was cleaned with deionized water, acetone, methanol, and isopropanol in ultrasonic bath with 10 min for each step. Then, the substrate was dried with hot air. The ITO electrode and Au wire were immersed in the precursor solution with spacing of 10 mm. The various voltages between 2.5 to 3.5 V were applied into two electrodes with a variable DC power supply. The deposition time was varied in duration of 60 to 120 seconds. Fig. 1 (a) depicts the setup of deposition system. Then, the deposited films were annealed at 450 °C for 2 hour in furnace. The properties of prepared films including crystalline structural characteristics, chemical bonding, and surface morphology, of prepared films were investigated by X-ray diffraction (SmartLab, Rigaku), FTIR (Spectrum Two FT-IR Spectrometer, PerkinElmer), and scanning electron microscope (EVO 15, Zeiss), respectively.

To fabricate the sensing layer of EGFET, the prepared CuO film was covered with polyimide tap with active area of 10 x 10 mm2. The EGFET system was composed of commercial N-MOSFET CD4007UBE, which the drain-source voltage (V_{DS}) and current (I_{DS}) can be performed with a source-meter (Keithley 2400). The gate of CD4007UBE was connected to the prepared CuO film. The Ag/AgCl was used as the reference electrode that the voltage at reference electrode can be controlled from precision voltage generator (Advantest R6144) defined as V_{GS} or V_{Ref}. The schematic of experiment setup are shown in Fig. 1(b)

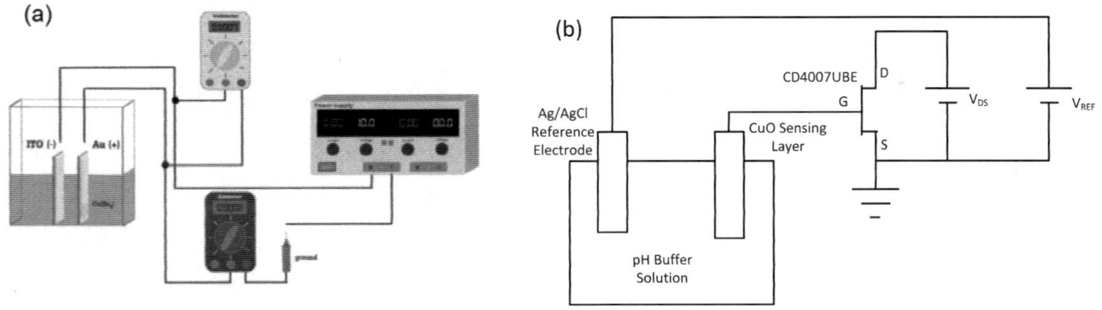

Figure 1. (a) The electrochemical deposition setup, and (b) the schematic measurement setup.

3. Results and Discussion

The crystalline structure of prepared films was performed with X-ray diffraction as shown in Fig. 2(a). The diffraction peaks were found at diffraction angle of 35.6, 36.2, 38.6, and 48.9 degree, which corresponded with orientation planes of (002), (111), (-111), and (-202) of CuO and Cu_2O phase structures [4]. The prepared films exhibited high crystallinity with increasing the deposition voltages and deposition.

Fig. 2 (b) depicts the FTIR spectra of prepared copper oxide films. The FTIR spectra can be used to confirm the functional groups in prepared copper oxide films. It can be observed the strong peaks located at wavenumber of 505 cm^{-1} which defined as Cu-O stretching [5]. The results obtained from the FTIR measurements are consistent with the results obtained from the XRD measurements.

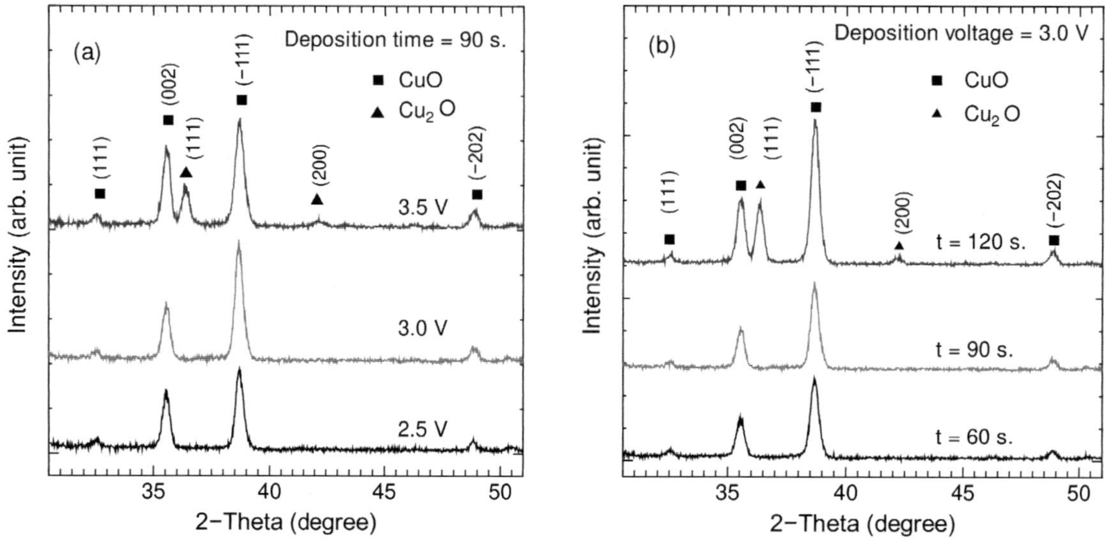

Figure 2. XRD spectra of (a) various deposition voltages, (b) various deposition times of CuO prepared by electrochemical deposition technique.

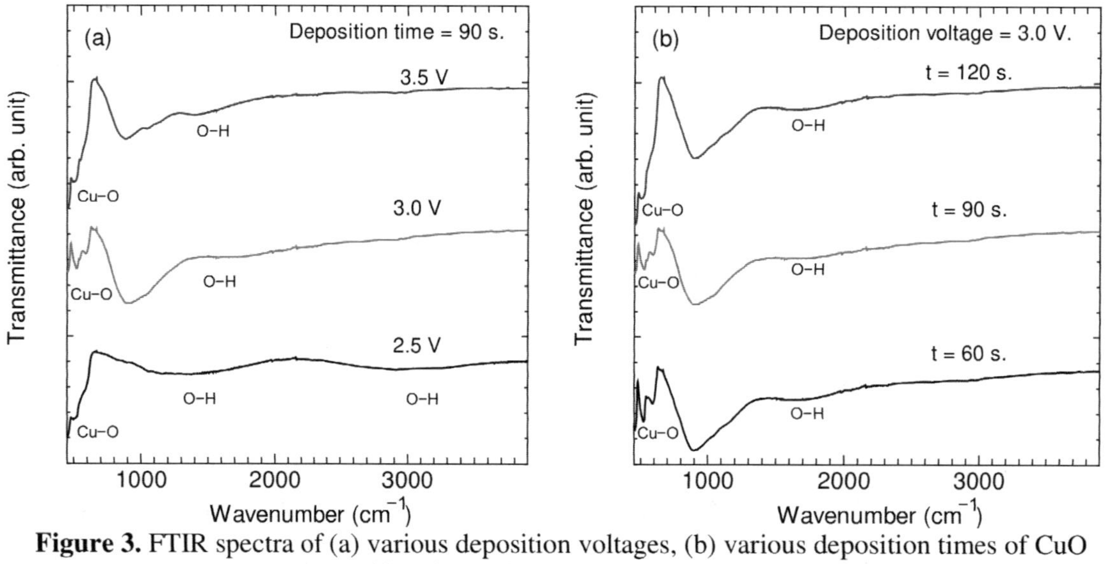

Figure 3. FTIR spectra of (a) various deposition voltages, (b) various deposition times of CuO prepared by electrochemical deposition technique.

Surface morphology of CuO films at different conditions of deposited voltages and deposited times are shown in Fig. 4. It can be seen the morphologies of films exhibited rough surface depended on preparation conditions. The cross-section images of prepared films are shown in the inset of each figure. The thicknesses of deposited CuO films were directly estimated from those images with ten points. The thicknesses of values of deposited films with deposition voltages of 2.5 V, 3.0 V, and 3.5 V with fixed deposition time of 90 second were 1475.2, 1520.2, and 2100.1 nm. In case of deposited films with fixed deposition voltage of 3.0 V, the thicknesses were 1405.6, 1520.2, and 3893.1 nm for

IOP Conf. Series: Materials Science and Engineering 1286 (2023) 012010 doi:10.1088/1757-899X/1286/1/012010

deposited films with deposition times of 60, 90, and 120 seconds. It can be seen that the thickness of CuO films increased with increasing of both deposition time and applied voltages due to the influence of higher potential and longer deposited time on the prepared films.

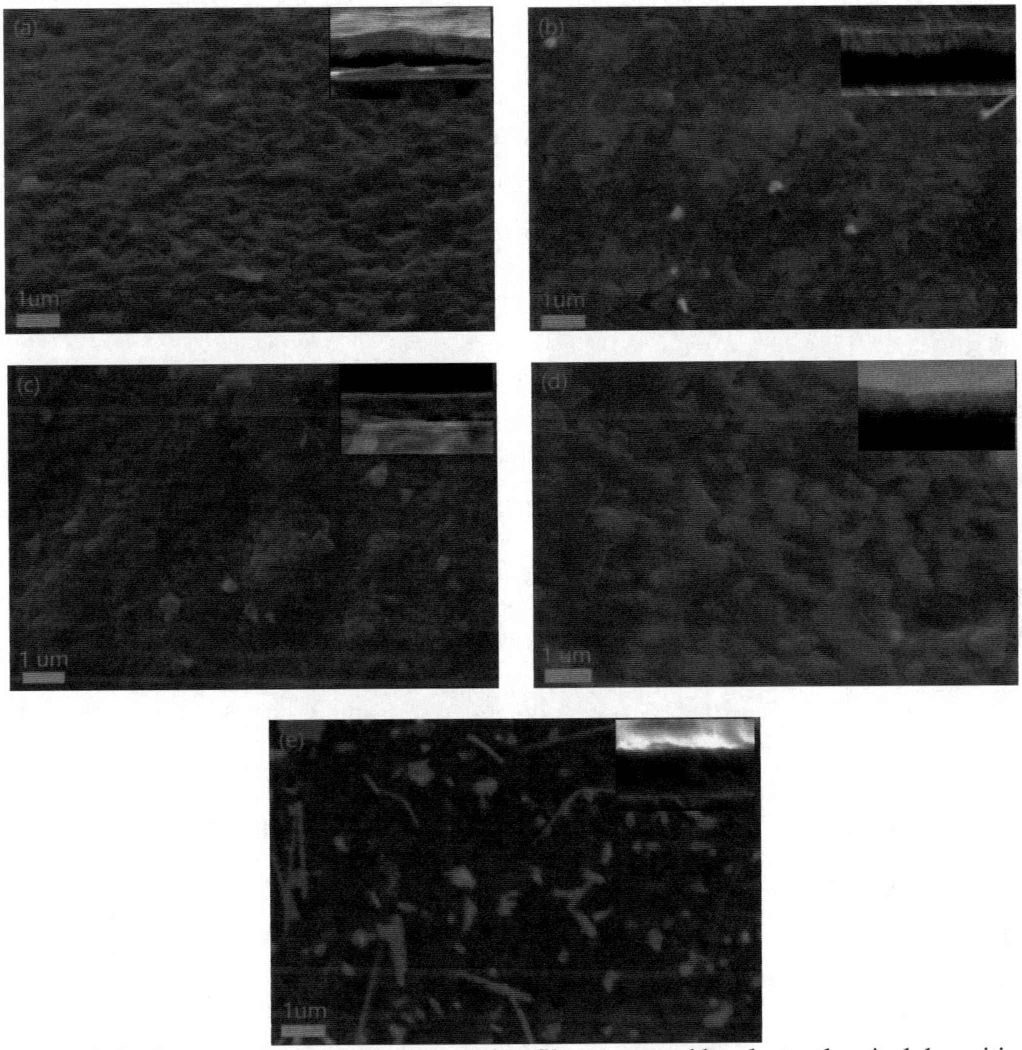

Figure 4. Surface morphology of copper oxide films prepared by electrochemical deposition technique at conditions of deposited voltage and deposited time: (a) 2.5V 90 second, (b) 3.0 V 90 second, (c) 3.5 V 90 second, (d) 3.0V 60 second, (e) 3.0 V 120 second. Inset of each figure showed the cross-section images of prepared films.

The prepared copper oxide films was used as sensing layer for extended-gate filed effect transistor with measurement setup as shown in Fig. 1(b). The relation of drain-source voltage (V_{DS}) and drain-source current (I_{DS}) at various pH levels are shown in Fig. 5 (a)-5(c) and Fig. 6(a)-6(c) with fixed he reference voltage (V_{REF}) at +3.0 V. It was found that the I_{DS} decreased with increasing the pH values. In addition, the relation of reference voltage (V_{REF}) and drain-source current (I_{DS}) at various pH levels are shown in Fig. 5 (d)-5(f) and Fig. 6(d)-6(f) with fixed V_{DS} at +0.3 V. It can be seen that the characteristics of I_{DS} and V_{REF} was shifted with increasing the pH value. Therefore, the current-voltage relation of EGFET with CuO film demonstrated the pH sensing. Which the relation of V_{DS}-I_{DS} can be used to calculate the pH current sensitivity [6] from

IOP Conf. Series: Materials Science and Engineering 1286 (2023) 012010 doi:10.1088/1757-899X/1286/1/012010

$$pH\ current\ sensitivity = \frac{\Delta\sqrt{I_{DS}}}{\Delta pH} \qquad (1)$$

And the relation of V_{REF}-I_{DS} can be used to calculate the pH voltage sensitivity [6] from

$$pH\ voltage\ sensitivity = \frac{\Delta V_{T(EGFET)}}{\Delta pH} \qquad (2)$$

The calculated sensitivity both current and voltage of EGFET device with CuO films at various deposition voltages and deposition times with the correlation coefficient are shown in table 1. Generally, the pH sensing characteristics on metal oxide material might be due to the changing of surface potential of pH buffer solution over the EGFET sensing layer, which can be described by the site-binding model theory [7].

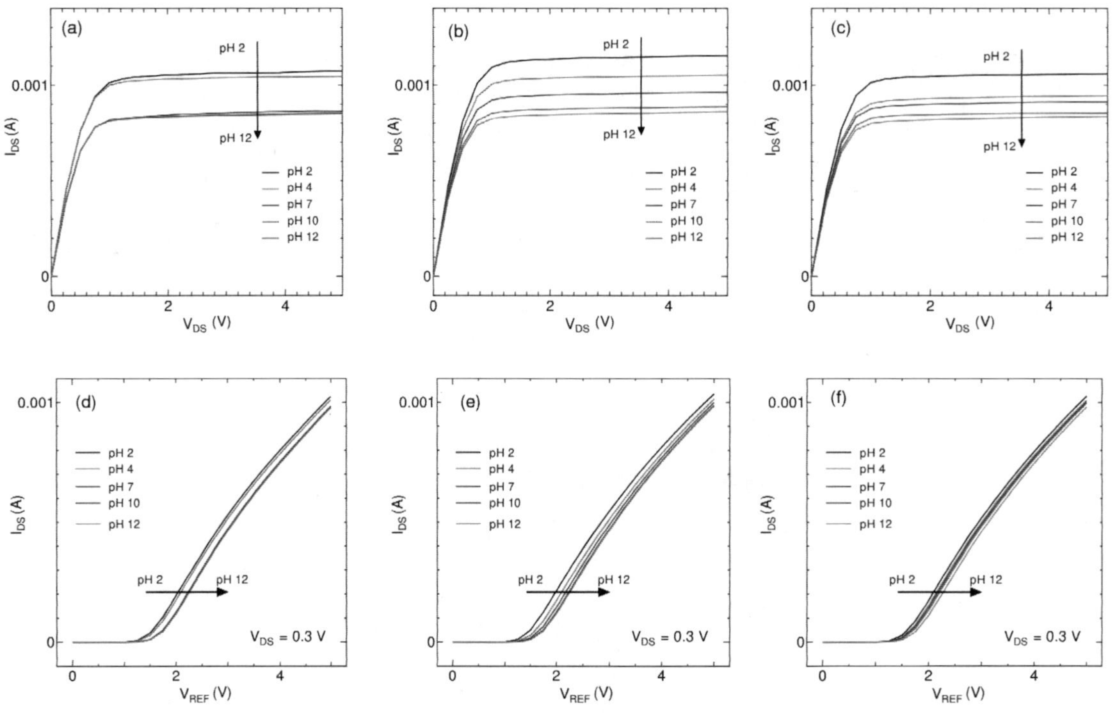

Figure 5. The relations of V_{DS} (drain-source voltage)-I_{DS} (drain-source current) and V_{RE}F(reference voltage)- I_{DS} (drain-source current) of CuO films prepared at varied deposition voltage of (a),(d) 2.5 V; (b), (e) 3.0 V, and (c),(f) 3.5 V, with fixed deposition time of 90 second.

IOP Conf. Series: Materials Science and Engineering 1286 (2023) 012010 doi:10.1088/1757-899X/1286/1/012010

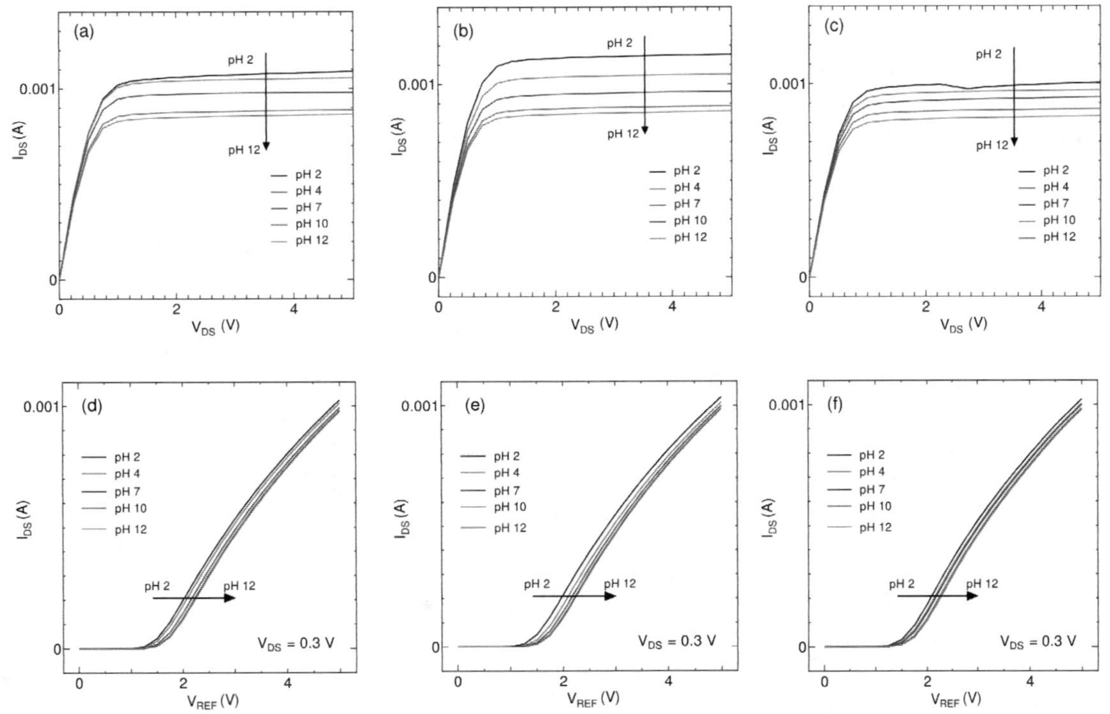

Figure 6. The relations of V_{DS} (drain-source voltage)-I_{DS} (drain-source current) and V_{REF}(reference voltage)- I_{DS} (drain-source current) of CuO films prepared at varied deposition time of (a),(d) 60 second; (b), (e) 90 second , and (c),(f) 120 second with fixed deposition voltage of 3.0 V.

Table 1. Summarised performances of prepared CuO EGFET.

Preparation conditions		Current sensing		Voltage sensing	
Applied voltage (V)	Deposition times (s)	Sensitivity x10^{-1} $(\mu A)^{1/2}$/pH	R^2	Sensitivity (mV/pH)	R^2
2.5	90	4.06	0.839	25.2	0.875
3	90	4.56	0.976	28.5	0.901
3.5	90	1.82	0.866	17.6	0.940
3	60	3.72	0.983	25.6	0.973
3	120	2.61	0.984	20.3	0.968

4. Conclusion

The electrochemical deposition technique with two electrode configuration was applied to prepare copper oxide films on commercial indium-tin oxide/glass substrate following with annealing at 450 °C in furnace. The CuO films were used to fabricate the pH sensor based on extended-gate field effect transistor (EGFET) with pH range of 2-12. It was found that the device demonstrated the highest voltage and current sensitivities of 28.5 mV/pH, and 4.56 $(\mu A)^{1/2}$/pH for CuO prepared at 3.0 V for 90 second. It can be seen that the prepared CuO films can be used as pH sensing application with EGFET device.

Acknowledgement

This work was financially supported by King Mongkut's Institute of Technology Ladkrabang (KMITL grant number: RE-KRIS/FF65/39) research fund. The authors acknowledge the facilities and technical assistance from Nanotechnology and Materials Analytical Instrument Service Unit (NMIS) of College of Materials Innovation and Technology, King Mongkut's Institute of Technology Ladkrabang.

Declaration of Interest

The authors declare no conflict of interest.

References

[1] Vinoth V, Shergilin T D, Asiri A M, Wu J J and Anandan S 2018 Facile synthesis of copper oxide microflowers for nonenzymatic glucose sensor applications *Materials Science in Semiconductor Processing* **82** 31-8

[2] Sabah F A, Ahmed N M, Hassan Z and Almessiere M A 2017 Influences of substrate type on the pH sensitivity of CuS thin films EGFET prepared by spray pyrolysis deposition *Materials Science in Semiconductor Processing* **63** 269-78

[3] Ahmed N M, Kabaa E, Jaafar M and Omar A 2017 Characteristics of extended-gate field-effect transistor (EGFET) based on porous n-type (111) silicon for use in pH sensors *Journal of electronic Materials* **46** 5804-13

[4] Kumar K, Priya A, Arun A, Hait S and Chowdhury A 2019 Antibacterial and natural room-light driven photocatalytic activities of CuO nanorods *Materials Chemistry and Physics* **226** 106-12

[5] Gund G S, Dubal D P, Dhawale D S, Shinde S S and Lokhande C D 2013 Porous CuO nanosheet clusters prepared by a surfactant assisted hydrothermal method for high performance supercapacitors *RSC advances* **3** 24099-107

[6] Al-Hardan N H, Abdul Hamid M A, Ahmed N M, Jalar A, Shamsudin R, Othman N K, Kar Keng L, Chiu W and Al-Rawi H N 2016 High sensitivity pH sensor based on porous silicon (PSi) extended gate field-effect transistor *Sensors* **16** 839

[7] Fung C D, Cheung P W and Ko W H 1986 A generalized theory of an electrolyte-insulator-semiconductor field-effect transistor *IEEE Transactions on Electron Devices* **33** 8-18

AUTHOR INDEX